A Short History

of

Botany in the United States

edited by

Joseph Ewan

with contributions by

CHESTER A. ARNOLD, KENNETH F. BAKER,
SHERWIN CARLQUIST, CHARLES B. HEISER, JR.,
STERLING B. HENDRICKS, GEORGE H. M. LAWRENCE,
EMANUEL D. RUDOLPH, PAUL B. SEARS, JERRY STANNARD,
WILLIAM RANDOLPH TAYLOR, ROLLA TRYON,
AND CONWAY ZIRKLE

HAFNER PUBLISHING COMPANY

NEW YORK AND LONDON

1969

Printed and published

by

HAFNER PUBLISHING COMPANY, INC.
31 East 10th Street
New York, N. Y. 10003

Library of Congress Catalog Card Number: 74-75143

Printed in U.S.A. by
NOBLE OFFSET PRINTERS, INC.
NEW YORK 3, N. Y.

President's Preface

by

Kenneth V. Thimann

It is an honor to introduce this book to the members of the XI International Botanical Congress, and to the scientific public. There is, of course, no "American Botany," any more than there is an American Science of any sort, for almost every branch of science is today too closely integrated throughout the civilized world to allow of the maintenance of really national schools. Examples of this integration appear throughout the book. There are, nevertheless, local centers of activity, like that at Harvard in comparative morphology and morphogenesis, or that until recently at Berkeley in trace elements and ion uptake. Indeed, the number and strength of these underlie the great vitality of the plant sciences in the U.S.A. now. We have at least five national societies in the general field of pure and applied botany with memberships of around 2000 or more—the botanists, the plant physiologists, phytopathologists, horticulturists and agronomists. Numerous smaller societies, plus the botanical membership of groups with broader interests, like those devoted to genetics or cell physiology, bring the total number up to what is probably the largest population of botanists in any country of the world. Thus if there is no "American Botany" there are plenty of American botanists.

So large a representation entails correspondingly large responsibilities. Some of these we can feel we have met, but others remain to be assumed. At least in recent years we have made it possible for many plant scientists the world over to come to America to participate in research or for specialized training. One can only hope that the current reduction in governmental research support will not foreshadow the end of this era, for it has been tremendously fruitful; the number of European, Japanese, Australian and Indian names as authors of research papers emanating from laboratories in this country bears witness to that.

We have finally—after a lapse of more than 40 years—brought the International Botanical Congress to this country again, and as hosts we shall try to make our guests feel welcomed, entertained and perhaps instructed. This book with its concise and scholarly presentation of past performance—often, alas, relatively slow and recent in development—should be a contribution to that last desire.

The responsibility that weighs most heavily upon some of us is that for helping to solve the food problem of the world. This will be a gigantic undertaking to which every aspect of botany from plant exploration to photo-synthesis, from taxonomy to molecular genetics, must contribute. Both applied and basic research will have important parts to play in this tremendous effort. For we shall need advances as great as the introduction of the plow, the use of nitrogenous fertilizers, or the understanding of pollination. The size of the world problem and its urgency will be in all of our minds during the Congress.

One of the main purposes of a historical review is to help us to see our successes, our failures, and our present problems in perspective. Professor Ewan and his co-authors have done much to this end, particularly by bringing the history down to the present day so that it becomes a part of our current research background. Perhaps an equally important purpose is to give us inspiration from the accounts of the great achievements of our predecessors. Such inspiration is the function of the Congress too. Its success, and that of this book, will be measured by the degree to which we are all inspired to continue and intensify our botanical work.

KENNETH V. THIMANN

vi

Preface

Botany is now such a large subject and the approaches to its study have become so diverse that no man can now claim to be a true 'generalist'. But all botanists are the better for seeing their specialties in the context of plant science in general. Similarly, Botanical Congresses have to be broken up into sections whose concerns are with very different aspects of the subject, but some of these sections offer better than usual opportunities for synthesis. In the XI International Botanical Congress, at Seattle, Section 9, dealing with the History of Botany, is one of these. Consequently, it is very satisfying for those of us in this Section to give our backing to *A Short History of Botany in the United States.*

Professor Joseph Ewan is a most highly qualified person to organize and edit this work and the twelve authors he has brought together are authorities in their fields. They have addressed themselves to an enormous task of describing not only the facts of history but also the circumstances in which they became factual. To their efforts in summarization and distillation we owe a considerable debt which we can repay by conscious efforts to make use of the information so that all may learn and appreciate the past, the better to plan the future.

HERBERT G. BAKER
Chairman, Section 9
Subcommittee of the Program Committee
XI International Botanical Congress
Seattle, 1969

Science is essentially international or suprana-
tional. There are no scientific problems which
are exclusively American. There is no American
science, but there are American scientists, a
good many of them, and some of them as great
as may be met anywhere else in the world.

George Sarton

Contents

Editor's Preface

In the tradition of the former International Botanical Congresses it was proposed by the Historical Section of the Botanical Society of America, Dr. Jerry Stannard in particular, that a Short History be prepared for the XI Congress. After financial arrangements were eased by the Hafner Publishing Company through the enthusiasm of Harry Lubrecht for the book, the thirteen areas to be reviewed were decided from the suggestions of many botanists. To all the authors I tender my sincere appreciation for their understanding and cooperation.

The list of "Selected Readings" is just that. Dates and bibliographic clues are provided in the text but no attempt has been made to give references in full. The bibliographies provided by the books and papers selected for the "Readings" should be consulted.

Your additions, corrections, and excisions are solicited, especially in respect to the Calendar of Events wherein important items may have been overlooked, and dates, persons, and places erroneously reported. Disagreement in the published and remembered record is incredible except to those who have pursued such facts.

I am grateful to Herman Hochschwender of the Tulane School of Architecture for preparing the tables.

Finally, this Short History could not have been finished without the head, heart, and hands of my wife, Nesta, who prepared the indices and assisted in all stages of its composition.

JOSEPH EWAN

Calendar of Events

To my knowledge this is the first chronology of this scope. Representative selected items are included to characterize the trends in the growth of the plant sciences. With the plethora of detail to be reviewed, the selection inevitably remains personal. If a bias toward systematic botany is noticed, it must be allowed that that fundamental subject predominated for over three hundred years. Full names of persons with vital dates are included in the index.

c. 300 B.C.

Storage baskets made of yucca leaves, juniper berry skewers, and wooden dice in use by natives who lived near Tularosa Cave, New Mexico.

900 A.D.

Seeds of *Lithospermum ruderale,* used by the American Indian in modern times as a contraceptive, stored by natives who lived at Lodaiska site near Denver.

1542

"Black Drink" from *Ilex vomitoria* by Indians of southeastern U. S. first described by Cabeça de Vaca in his *Relacion y Commentarios.*

1562

Ribaut and Laudonniere anchored at the mouth of the "Riviere de May" in "Florida."

1569

Monardes described Sassafras given him by that name by a Frenchman from "Florida."

1576

Sarracenia purpurea mentioned by Clusius under the name "Limonio congener," with figure wanting flower, was based on a plant furnished by a Paris apothecary, Claude Gonier. He had obtained it from Lisbon whither it may have been taken by a Portuguese sailor out of Newfoundland.

Sarracenia flava described by de Lobel in his *Nova stirpium adversaria.*

1588

Hariot's *A Briefe and True Report of the New Found Land of Virginia* (London) is published, the first book in vernacular English devoted to the flora and fauna of what is now the U. S.

1586

John White's drawings of Virginia plants (mostly lost?) from Roanoke Island are the first from present U.S.

1600

By this year thirty species of plants from the U. S. and Canada had been reported as growing in European gardens (see Table III).

1612

Strachey returned to England from Virginia. His "Historie," which was suppressed as commercially unfavorable, contained plant references.

1634

Tradescant's *Catalogus* listed forty species from Virginia.

1635

Cornut's *Canadensium plantarum historia,* with proficient illustrations of North American plants, published in Paris.

1647

Gov. Stuyvesant of New Amsterdam planted a Summer Bonchretien, a grafted pear which lived for over two centuries at his garden, in the *bouwery.*

1672

Josselyn lists 23 "such Plants as have sprung up since the English Planted and kept cattle in New England."

1678

Banister arrived in Virginia. For the following fourteen years he sent seeds, specimens, drawings, and notes, toward a "Natural History" to English correspondents, until his death in 1692.

1683

William Penn distinguished five oaks and ten other trees of Pennsylvania.

1698

Edward Shippen, Quaker merchant, had an "extraordinary fine and large garden" in Philadelphia.

1713

400 American plants were known in Britain by this year of Bishop Compton's death. He had been active in their introduction.

1716

First unambiguous account of plant hybridization is recorded by Cotton Mather: it involved red and blue kernels of *Zea mays.*

1720

Dudley published on maple sugar in *Philosophical Transactions.*

1728

John Bartram began assembling native plants at his farm near Philadelphia.

1731

Library Company of Philadelphia was founded by Franklin who made
  arrangements for importation of English books on gardening, etc.

1735

Experimental gardens for testing subtropical crop plants established
  near Savannah, Georgia.

1737

Robert Prince established a commercial nursery of eight acres at
  Flushing, Long Island, which would be in business for 130 years.

1739

Gronovius published part one of *Flora Virginica* in Leiden from
  Clayton's notes and specimens.

1747

Redwood Library founded in Newport, R. I., by Abraham Redwood,
  merchant and philanthropist, who developed a private botanical
  garden with hothouses at nearby Portsmouth.

1749

November 30. Peter Kalm visited Bartram's garden.

1755

Alexander Garden, resident of Charleston, wrote his first letter to
  Linnaeus.

1756

Bolzius published an account of 80 wild and cultivated plants growing
  at Ebenezer Colony near Savannah, Georgia, in *Hamburges
  Magazin*.

1760

Gov. Arthur Dobbs discovered Venus Fly-trap in North Carolina and
  sent a description to Collinson.
John Bartram visited Alexander Garden at Charleston.

1764

John Bartram sent his Carolina journal of 1762 to Collinson asking
  that Solander be invited to peruse it.

1765

October. *Franklinia* (past flower) discovered by John and William
  Bartram on the Altamaha River, near Fort Barrington, Georgia.
DuSimitière arrived in the U. S. with a plan for the first natural history
  museum at Philadelphia.

1767

William Young, Jr., botanical poseur, collected 302 species of Carolina
  plants and made crude drawings in color.

1768

Adam Kuhn, native of Germantown, Pa., student of Linnaeus in 1761, became professor of Materia Medica and Botany in the College of Philadelphia, the first professor of botany in the colonies.

First professional gardener in the colonies, James Alexander, was elected to the American Society (i.e. American Philosophical Society).

1770

William Bartram began drawing plants and animals for Dr. Fothergill.

1772

Fothergill proposed a natural history expedition to Florida to William Bartram.

1773

Charleston Museum opened with Pinckney as its first curator.

Humphry Marshall established the second botanic garden in colonial Pennsylvania at West Bradford, Chester County.

1774

William Bartram sent about 100 "very curious" dried specimens to Fothergill.

1775

Battle of Lexington opened the Revolutionary War.

1777

September 22. John Bartram died, aged 78, and his son, John Bartram Jr., became proprietor of his garden.

1778

January 18. Capt. Cook, aboard H. B. M. *Resolution,* arrived at the Sandwich Islands on his third voyage to the South Seas. David Nelson made the first collections on Oahu.

1784

Waterhouse, who as a lad read at the Redwood Library, offered public lectures in botany and mineralogy at Rhode Island College while a member of the Harvard Medical School faculty.

Jefferson arranged with Thouin in Paris for an exchange of seeds and plants.

Landreth opened the first seed business in the U. S. in Philadelphia.

1785

Philadelphia Society for Promoting Agriculture founded.

Marshall's *Arbustum Americanum; the American Grove* published in Philadelphia, the first American imprint devoted expressly to botany.

1787

André Michaux made his first botanical foray through the Carolinas and established his second growing garden about ten miles from Charleston, at this time the largest city in the South.

1789

Ginkgo was planted at Pierce Arboretum, Kennett Square, Pa. (see 1968).

1788

Walter's *Flora Caroliniana,* which included Walter's local Santee River and Fraser's piedmont Carolina plants, published in London.

1791

William Bartram's *Travels* published in Philadelphia. It included many undescribed plants of southeastern U. S., some illustrated.

Sandalwood trade began in the Sandwich Islands with the sloop *Lady Washington,* Capt. Kendrick.

1792

Mitchill appointed first professor of agriculture (and botany) at Columbia College, New York.

1801

Hosack established his Elgin Botanic Garden of twenty acres including native plants and by exchange exotic species from London, Copenhagen, Paris, Florence, and St. Vincent, British West Indies.

1802

M'Mahon, Philadelphia nurseryman, who had arrived in 1796, issued a one-page broadside of nursery items. He later sold novelties brought back from the Pacific Northwest by the Lewis and Clark Expedition.

1803

Michaux's *Flora boreali-Americana,* the first flora of national scope, published in Paris, with many plates by Redouté.

Lyon reported seeing "6 or 8 full grown" *Franklinia* trees in the wild —the last time seen—at the same locality where the Bartrams had discovered them.

1804

Soybeans first brought to the U. S. as ballast.

1805

Philadelphia scientists and naturalists honored the visitors Humboldt and Bonpland at a dinner at Peale's Museum.

1806

Hosack published the first edition of his *Catalogue* of the plants grown at the Elgin Botanic Garden.

1807

Harvard Botanic Garden established in Cambridge by Professor Peck with encouragement from John Lowell.

John Bartram, Jr., issued a 33-page *Catalogue* of plants for sale at Bartrams' garden.

1808

Nuttall arrived in Philadelphia and met Professor Barton who engaged him to collect plants towards a comprehensive flora.

Lafon published Louisiana's first mycological paper on *Cordyceps* growing on larva.

1811

Waterhouse published *The Botanist,* an anthology of essays on philosophical botany which he had delivered over the past 25 years at Harvard and Rhode Island College.

Drowne elected Professor of Materia Medica and Botany at Brown University.

1812

March 17. Academy of Natural Sciences founded in Philadelphia. It was destined to receive the nation's oldest plant collections and to assemble one of its foremost botanical libraries.

1813

Literary and Philosophical Society of South Carolina established with important influence of Elliott. Poinsett was an active member.

1814

Pursh's *Flora Americae septentrionalis,* the first flora on a continental scope and that included Pacific Northwest species, published in London.

First local flora of New England, Bigelow's *Florula Bostoniensis,* using the Linnaean system, published in Boston.

1815

Jefferson sold his 7000-volume library to the Nation as a nucleus for the Library of Congress. He owned 26 botanical titles.

Correa da Serra presented the new Jussieu's Natural System of Classification in botany lectures at Philadelphia.

1817

Gov. Clinton and Dr. Hosack were elected the first American members of what was to become the Royal Horticultural Society.

Coxe, "father of American pomology," published *A View of the Cultivation of Fruit Trees* in Philadelphia.

1818

Dr. Daniel Drake offered the first instruction in botany west of the Alleghenies as a "course of Botanical Lectures" in Cincinnati.

Nuttall published his *Genera* based on his botanizing excursions of the past decade. He followed the Linnaean Sexual System as a matter of convenience and in deference to the then current instruction.

Schweinitz published on fungi of North Carolina, in Latin, enumerating over 1000 species, of which 300 had been undescribed.

Peck issued a *Catalogue* of Harvard Botanic Garden, to which "strangers of distinction, and clergymen" were admitted *gratis,* others by ticket.

1820

First U. S. *Pharmacopoeia* published.

1822

New York Horticultural Society, first of its kind in the United States, incorporated with Hosack as president.

1823

Nuttall took visiting David Douglas on a tour of Bartrams' garden, then in a "deplorable state."

1826

This the year of Jefferson's death opened with 25 scientific societies in the U. S., over half devoted to natural history, many of recent founding.

"Silkworm Mulberry" mania that was to last for fifteen years set off by Act of Congress to encourage planting.

1827

Pennsylvania Horticultural Society, the oldest society in continuous existence in the nation, organized in Philadelphia.

1829

Amherst College established the second institutional herbarium in the U. S.

1830

Boston Society of Natural History elected Nuttall its first president and founded its herbarium incorporating the C. J. Sprague and Thomas Taylor lichens and Francis Boott collections.

1833

Henry Perrine planted sisal and henequen on Florida Keys as part of his tropical plant introduction program.

First *U. S. Dispensatory* published.

1835

January 4. Nuttall arrived with Townsend in Honolulu on their first visit to the Hawaiian Islands.

1836

Asa Gray published his first book, *Elements of Botany,* in New York.
Massachusetts Horticultural Society founded in Boston.

1838

Lt. Charles Wilkes sailed with a fleet of six naval vessels staffed with
nine "scientific gentlemen" on our first government-sponsored
expedition around the world.

Dreer opened his seed business in Philadelphia, featuring ornamental
trees, flower and vegetable seeds.

1839

Asa Gray toured European herbaria checking type specimens for the
Flora of North America on which he was collaborating with
Torrey.

1840

National Institution for the Promotion of Science organized on May
15th at Poinsett's house in Washington with eight persons present.

Amos Eaton published the 8th and last edition of his *Manual,* and the
curtain fell in this country on the Sexual System of Linnaeus.
Some 18,000 copies of all editions had been printed.

Blodgett reported for the first time some twenty West Indian plants
growing on the Florida Keys.

Rafinesque, author of 6700 binomials (including 2700 proposed generic
names) died of cancer, aged 57, in a Philadelphia garret.

1841

Brackenridge collected the endemic insectivorous *Darlingtonia* on
Mount Shasta—the "most dramatic discovery of the Wilkes
Expedition."

1842

Gray published the first edition of his *Botanical Text-Book.*

April. Tuckerman of Amherst College bought at auction in London
some lots of Lambert's books and collections.

Wilkes Expedition returned in June having criss-crossed the Pacific,
mapped whaling grounds, observed and described aborigines, and
collected plants (9,674 species) and animals.

1845

Wood's *Class-Book of Botany,* Gray's chief competitor, first appeared
in Boston.

1847

First American-made microscope with an efficient achromatic lens combination produced by Spencer.

1848

May 4. Jane Lathrop Loring became the wife of Asa Gray.

1849

Kellogg, who had shipped around the Horn, arrived in Sacramento, California, and lived to be the state's first resident botanist.

1850

March. American Association for the Advancement of Science met in Charleston. Visiting Professor Harvey of Dublin discussed marine algae.

1851

Hunnewell's private arboretum established adjacent to the campus at Wellesley, Mass., became one of the oldest and finest conifer and *Rhododendron* collections in the East.

1853

Christmas Eve. Lindley in the *Gardener's Chronicle* proposed *Wellingtonia gigantea* for the Sierra Redwood based on Lobb's specimens collected in the Calaveras Grove.

1854

"Mammoth Tree" exhibited in New York City from reconstruction of a 60-foot bark section stripped from a *Sequoiodendron* over 300 feet tall which had grown in the Calaveras Grove of California.

Warehouse fire destroyed all but 24 copies of Brackenridge's *Filices* of the Wilkes Expedition, rendering it one of the rarest of American botanical works.

1860

"Mr. Shaw's Garden" opened to the public in St. Louis.

D. C. Eaton took the Ph.D. degree at Harvard submitting the thesis "Filices Wrightianae et Fendlerianae" involving 360 fern species.

1861

Engelmann published the first paper on plant pathology in the U. S.

1862

President Lincoln signed the Morrill Act which distributed 13,000,000 acres of public domain to the states (area = Belgium) for establishment and maintainance of land-grant colleges, favoring "the agricultural and mechanic arts." Botanical courses were prominent in their curricula.

1863

National Academy of Sciences incorporated. Asa Gray was a founding member.

Porcher's *Resources of Southern Fields and Forests* published to supply Confederate troops with what in the 20th Century wars came to be called "survival" information.

1864

Lloyd Library of Cincinnati founded, notable for its florulas as well as for pharmaceutical botany.

1867

J. M. Bigelow recommended to the American Medical Association that cinchona be introduced and cultivated in the U. S.

Parry founded herbarium at the Public Museum in the frontier town of Davenport, Iowa.

1868

Hoopes' *Book of Evergreens,* the first treatise on native conifers, published.

Parry appointed first curator of the National Herbarium which was transferred from the Smithsonian Institution to the Department of Agriculture.

1869

Burrill founded the herbarium at University of Illinois.

1870

Berthoud became first professor of botany in Colorado at the School of Mines at Golden. Nearby Denver had been founded eleven years before.

1871

Dr. Hillebrand, author of the *Flora of the Hawaiian Islands* (1888), left after twenty years residence at Honolulu. His garden of exotic plants formed the nucleus of the present Foster Botanical Garden.

Bessey added laboratory work to his botany course at Iowa Agricultural College, Ames.

President A. D. White purchased the collection of Horace Mann thus establishing the Cornell University herbarium.

1872

Tuckerman's *Genera Lichenum* published.

Goodale, student of Tuckerman at Amherst, began instructing botany at Harvard under Asa Gray.

1873

Navel oranges introduced from Brazil in 1870 by William Saunders for the U. S. Department of Agriculture, planted at Riverside, California, by Jonathan and Eliza C. Tibbetts, were used as grafting stock for the industry.

1874

Farlow opened the first laboratory devoted to cryptogamic botany at Harvard.

Botanical Gazette—a monthly of four pages—founded by John and Stanley Coulter.

1875

Apgar, whose *Plant Analysis, adapted to Gray's botanies,* (1st print. 1874) became the familiar work-book of botany classes, founded the herbarium at State Teachers' College, Trenton, N. J.

W. O. Atwater, co-inventor of the colorimeter, appointed director of the country's first state agricultural station at Middletown, Connecticut.

First plant pathology course taught by Farlow at Harvard.

1876

First course in economic botany, later to be called "Plants and Human Affairs," begun at Harvard.

1877

Shortia, lost since its discovery by Michaux in 1787, rediscovered by Hyams on Catawba River, McDowell Co., N. C.

Gray and J. D. Hooker traveled together through Colorado and Utah to California on a phytogeographic tour.

First edition of *Naturalist's Directory* published by Samuel Cassino in Salem, Massachusetts, facilitating exchanges of collections and fostering contacts among amateurs and professional botanists: 38th ed. (1958).

Botanic garden founded at East Lansing, the oldest in Michigan still in continuous operations at the same site. First directed by W. J. Beal who served for 33 years.

First fascicle of D. C. Eaton's *Ferns of North America* published.

1878

Arnold Arboretum founded by Sargent.

Luther Burbank moved from Lunenburg, Mass., to Santa Rosa, California, to continue plant breeding.

1879

Burrill, working on apple and pear blight, first attributed a plant disease to bacterial origin.

Lesquereux published the first volume of his *Coal Flora.*

Beal began long-term longevity experiments using seeds of 20 weed species.

1880

C. V. Piper and members of the Young Naturalists Society of Seattle, started an herbarium which became that of the University of Washington. Ten years later Piper founded the herbarium at State College, Pullman.

1881

Loganberry produced by James Harvey Logan, jurist and horticulturist, in his garden in Santa Cruz County, California.

1882

Engelmann published the first American generic monograph of Pteridophyta—on the genus *Isoetes.*

Cyclone partially destroyed Prof. Bessey's botanical laboratory at the University of Nebraska but a student rescued his one compound microscope from the wreckage.

Lemmon and his wife botanized in Arizona—Mount Lemmon commemorates them—and erroneously reported the discovery of a form of the original South American *Solanum tuberosum.*

1884

Sargent published the first comprehensive synopsis of North American trees in Volume IX of the Tenth Census.

Wolle published the first of four then widely used volumes on algae.

The book Sullivant had hoped to write, the *Manual of North American Mosses,* published by Lesquereux and James.

Cotton Centennial Exposition opened in New Orleans including plant exhibits from Texas and Louisiana by Joor, and California forest trees arranged by Lemmon and his wife.

Pharmaceutical firm of Parke, Davis & Co. employed Rusby, a systematic botanist, to search for, collect, and verify drug plants— one of the first such appointments.

1885

W. A. Kellerman, aged 35, founded the *Journal of Mycology,* which he edited intermittently for sixteen years, and which became *Mycologia* after his death in Guatemala in 1908.

Jesup established a comprehensive collection of North American woods at the American Museum of Natural History, giving the Museum a bequest of $2,000,000.

Goodale, a pupil of Pfeffer, published the first "physiological botany" in this country as Volume II of Gray's *Botanical Text-Book.*

November 18. Birthday greetings sent to Asa Gray by 180 botanists from around the world.

1886

A. R. Wallace visited the U. S. on a lecture tour.

1888

Dawson's *Geological History of Plants,* the first paleobotanical text by a North American author, published.

1889

Herbarium organized with the founding of the Bernice Pauahi Bishop Museum at Honolulu by C. R. Bishop.

1890

At Purdue University Arthur established what was to become one of the richest collections of rust fungi.

F. L. Harvey founded the herbarium at University of Maine at Orono. He established the University of Arkansas herbarium in 1905.

1891

Sargent published the first volume of his 14-volume *Silva of North America.*

Judge Addison Brown drafted the legislation enabling the establishment of the New York Botanical Garden.

Stanford University opened with Campbell and Dudley the first professors of botany.

1892

Sturtevant presented his notable prelinnean library to the Missouri Botanical Garden.

California vine disease cost $10,000,000 this year.

1893

L. H. Bailey published the first detailed study of growth of plants under artificial light.

Horticultural Congress of 150 delegates met in Horticultural Hall, at World Columbian Exposition, Chicago.

Field Columbian Museum, Chicago, established its herbarium destined to be outstanding for neotropical botany.

Glass flowers collection, the work of Leopold Blaschka and his son Rudolph, presented to Harvard University. The collection ultimately numbered 184 plant families, and is visited by 250,000 persons each year.

Linnaean Fern Chapter of the Agassiz Association organized by Clute.

1894

National Herbarium recalled from the Department of Agriculture to the Smithsonian.

1895

Setchell succeeded E. L. Greene who had been the first Professor of Botany at Berkeley.

J. K. Small's *Monograph of the North American species of the genus Polygonum,* described at the time as the "most sumptuous American botanical thesis ever published."

Harshberger introduced the word 'ethnobotany.'

1896

New York Botanical Garden established.

Setchell and Jepson botanized by horse and wagon from Berkeley to Santa Cruz Mts., across the San Joaquin Valley, to Yosemite and return.

1897

Plant World launched by Knowlton.

MacDougal awarded the Ph.D. degree from Leipzig *in absentia* for his thesis on "Curvature of roots."

1898

Office of Foreign Seed and Plant Introduction of the U. S. Department of Agriculture organized by Fairchild with assistance of Cook, Ricker, and Swingle.

Cowles, Coulter's first doctoral candidate at Chicago, through his thesis on the dynamics of sand dune vegetation, contributed to the founding of plant ecology in this country.

Wheat rust cost the U. S. $67,000,000 this year according to Galloway.

Barnes proposed the term 'photosynthesis.'

1899

Sturtevant published his "Varieties of Corn" based on endosperm characters.

1900

Yale School of Forestry founded.

Asa Gray elected to the Hall of Fame, as one of 29 original electees.

L. R. Jones launched a vigorous program of studies in plant diseases at University of Vermont.

New York State Experiment Station at Geneva established primarily for fruit breeding programs on a broad scale, carried on for ten years under Hedrick.

1901

Mohr, for over forty years a pharmacist at Mobile, published *Plant Life of Alabama.*

A. S. Hitchcock moved from Kansas State College to the U. S. Department of Agriculture to become the nation's leading agrostologist.

MacMillan established the Minnesota Seaside Laboratory at Port Renfrew about 65 miles west of Victoria, B. C., one of the early field stations on the Pacific Coast.

Swimming sperm in *Zamia* reported by H. J. Webber.

1902

Second International Conference on Hybridization and Plant Breeding.

57 members of the antecedant Botanical Society of America delivered 40 papers at its ninth annual meeting, leading Galloway to remark that "the number of papers presented was so great that it was

impossible for visiting botanists to take anything like advantage of them."

Jeffrey proposed term 'megaphyllous plants' for Pteropsida without leaf-gaps; 'microphyllous plants' for Lycopsida with leaf-gaps.

1903

T. J. Howell, self-taught Oregon pioneer, published the seventh and last fascicle, all hand set, of his *Flora of Northwest America.*

J. K. Small, aged 34, published his *Flora of the Southeastern United States,* of almost 1400 pages, two years after his first visit to Florida.

1904

Mann completed his classic *Report on the Diatoms of the Albatross Voyages in the Pacific Ocean.*

Chestnut blight from Japan detected in the New York City area.

Devastating epidemic of wheat rust set off research programs for control measures.

1905

North American Flora launched by the New York Botanical Garden.

Safford, formerly of the U. S. Navy, published his beachmark *Useful Plants of the Island of Guam.*

Railroad magnate Henry E. Huntington started his botanical garden at San Marino, California, with 200 acres devoted to palms, cacti, and sub-tropical ornamentals, superintended by William Hertrich.

Tung oil industry launched at Chico, California.

1906

April 18. San Francisco earthquake and fire destroyed most of the library and herbarium of the California Academy of Sciences. Alice Eastwood and volunteers rescued hundreds of type specimens.

Hugo de Vries visited the U. S., including a trip to North Carolina with Professor Macfarlane and a transcontinental botanical tour by train with frequent stopovers

Botanical Society of America founded by the merger of three antecedant organizations; its membership: 119.

Petrified Forest National Monument, of Triassic *Araucarioxylon* with the fern *Laccopteris,* etc., established in Arizona.

University of Idaho herbarium, including Henderson collections, destroyed by fire.

American Museum of Natural History selected ten men who had advanced science the most in America: Agassiz, Audubon, Baird, Cope, Dana, Franklin, Henry, Humboldt, Leidy, and one botanist, Torrey.

1907

First university department of plant pathology established at Cornell.

W. A. Kellerman opened a school of tropical botany in Guatemala.

"Chinese Wilson" made his third collecting trip, which was the first for the Arnold Arboretum, followed by another in 1910 and two to Japan, 1914 and 1918.

Garfield Park Conservatory—"easily one of the outstanding conservatories in the country if not in the world" (Wyman) —established in Chicago.

1908

Robinson and Fernald published a revision, the seventh edition, of Gray's *Manual,* which became the most widely used identification guide in U. S.

A. C. Crawford suggested obligate relationship between certain plants and seleniferous soils.

Bryophytes of Connecticut, one of the most thorough studies of its kind, published by Evans and Nichols.

American Phytopathological Society founded.

1909

Duggar's *Fungous Diseases of Plants,* the first text on the subject, published.

Livingston began teaching plant physiology at Johns Hopkins University. In the next 30 years he and his students published nearly 300 papers.

Bioecologist Shimek, who taught 46 years at the State University, opened Lakeside Biological Laboratory at Lake Okoboji, Iowa.

Green Algae of North America published by Collins. His life cycle studies lead to Gilbert Smith's later work on sexuality.

Theodore Roosevelt added over 100,000,000 acres to the National Forests during his second term as President.

Ramaley opened first field station for ecological studies in the Rocky Mountains at Tolland, Colo., ele. 7500 ft.

1910

This year the College of Hawaii (present University of Hawaii), opened two years before, had an enrollment of 19 students. MacCaughey from Cornell taught botany.

Smithsonian Institution sponsored a biological survey of the Canal Zone before construction of the Canal.

Boy Scouts of America, incorporated this year, fostered field botany with its "merit badge" given for forest tree identification.

Coville began experimenting with blueberries.

1911

Harshberger's *Phytogeographic Survey of North America* published.

1912

International Phytogeographical Excursion toured from coast to coast.
Hugo de Vries made his second visit to the botanic gardens of the U. S.

1913

Karl F. Kellerman launched *Journal of Agricultural Research.*
Plant quarantines first imposed on a national basis.

1914

First spring flower show on a national scale sponsored by the Horticultural Society of New York and the Florists' Club.

1916

National Research Council organized.

Nichols advised the American Red Cross on the use of sphagnum in
surgical dressings.

Calamitous epidemic of wheat rust destroyed nearly 300,000,000
bushels in the U. S.

Clements' *Plant Succession* published.

1917

Pennell and Rusby botanized in Colombia seeking stands of wild
quinine as part of the war effort. Their 4791 general collections
made this year initiated a 12-year cooperative field exploration of
the northern Andes by the New York Botanical Garden, Harvard
University, and the Smithsonian Institution, which resulted in
20,000 numbers by later collectors.

Barberry eradication program legalized in North Dakota as part of a
national campaign to check wheat rust and increase wheat production in a war year.

175,344,000 bushels of corn destroyed by plant diseases this year.

1918

Britton published his *Flora of Bermuda,* the first of a series of West
Indian accounts based on nearly 100,000 specimens assembled
over a period of 20 years at the New York Botanical Garden.

July 16. President Wilson ordered the Natural History Building of the
Smithsonian closed to the public and all the exhibition halls
turned over to the Bureau of War Risk Insurance for offices.

1919

Final fascicle issued of Setchell, Holden and Collins' *Phycotheca-
Boreali-Americana,* numbering some 200,000 algal specimens.

Farlow cryptogamic collections initiated in 1874 acquired by Harvard.

J. A. Harris published a pioneer biometric analysis of plant saps.

Diatom research laboratory established by Carnegie Institution of Washington at the Smithsonian through the influence of Mann.

1920

Garner and Allard published their classic paper on photoperiodic effects on growth and reproduction in plants.

First of five parts of Standley's comprehensive *Trees and Shrubs of Mexico* published.

200 *Eospermatopteris* stump casts of Devonian age discovered near Gilboa in the Catskill Mountains.

Setchell made the first of many journeys to Polynesia to study coral reef formation, ethnobotany, and patterns of plant distribution.

1922

Morton Arboretum of 735 acres dedicated at Lisle, Illinois, aiming to grow every woody plant that will live out of doors there.

During his pioneer tissue culture experiments W. J. Robbins reported that addition of yeast to the medium improved the growth of excised root tips of tomato.

Fossil Cycad National Monument (of cycadeoids) opened in South Dakota.

1923

Merrill moved to Berkeley as Dean of College of Agriculture after 22 years at Manila.

Noé discovered coal balls in Illinois.

1924

Holman and Robbins published the first edition of their *Textbook of General Botany*. It became the most widely used botany text ever adopted in college classes.

American Society of Plant Physiologists founded.

Boyce Thompson Institute for Plant Research founded at Yonkers, N. Y., and a desert botanic garden at Superior, Arizona.

Pioneer survey of hayfever plants published by allergists Duke and Durham for the Kansas City, Missouri, area.

Ring chromosome formation reported in *Oenothera* by Cleland.

1925

Clausen and Goodspeed first demonstrated experimentally Winge's hypothesis of origin of species by amphidiploidy using *Nicotiana*.

Bailey's *Standard Cyclopedia of Horticulture*, "a magnificent piece of book-making," long known in 6-volume edition, issued in popular 3-volume format.

Belling suggested heat treatment to induce polyploidy.

Jepson published his highly successful *Manual of Flowering Plants of California*.

Eddy Tree Breeding Station, now Institute of Forest Genetics, founded at Placerville, California, to breed pines for timber production.

1926

Crystallization of urease accomplished by J. B. Sumner.

August 16. L. H. Bailey opened the International Congress of Plant Sciences (Fourth International Botanical Congress) at Cornell University.

Rydberg debated with Skottsberg at the Ithaca Congress on the merits of recognizing large comprehensive genera vs. small narrowly defined genera.

Blaksley Botanic Garden (now Santa Barbara Botanic Garden), 30 acres largely devoted to native chaparral flora, founded in Mission Canyon, Santa Barbara, California.

Killip and A. C. Smith collected 7000 numbers of Colombian plants in five months' travels from Turbaco to the high paramos in the first intensive survey of Colombia's flora.

Henry A. Wallace advertised to the public hybrid corn seed that had been established experimentally in 1917.

1927

Rancho Santa Ana Botanic Garden founded at Anaheim, California, by Mrs. Susanna Bixby Bryant.

Santa Barbara Museum of Natural History herbarium with emphasis on the endemic California insular floras founded by ornithologist Ralph Hoffmann.

1928

Schaffner first suggested orthogenesis as important in angiosperm evolution.

Mammalogist Tate explored Cerro Duida on the Guiana-Venezuelan border. His collections of extraordinary endemic species and several new genera reported later by Gleason.

1929

California Botanic Garden, founded in 1927 in Mandeville Canyon near Beverly Hills, was a victim of the Depression. The Bonati Herbarium of 60,000 sheets purchased in Paris by Director Merrill passed to the University of California at Los Angeles.

Elgin Botanic Garden site dating from 1801 leased by Columbia University to Rockefeller Center at a rental of $3,000,000 a year.

Paleobotanist David White's *Flora of the Hermit Shale, Grand Canyon, Arizona*, incorporated correlative field surveys and extensive analyses of fossil collections.

1930

Hoagland of Berkeley given the Stephen Hale Award for his pioneer research on role of minor elements in plant nutrition.

Scarth and Lloyd published a successful *Elementary Course in General Physiology*.

Van Niel reported the parallel between photosynthetic processes in purple bacteria and higher plants.

University of California Botanic Garden (relocated over the mid-campus garden started by E. L. Greene) established in Strawberry Canyon above Berkeley.

1931

Creighton and McClintock using special stocks of *Zea mays* report that when chromosomes heteromorphic in two regions pair, parts are exchanged at the same time genes assigned to those regions are exchanged.

1932

Montreal Botanic Garden established.

Ynes Mexia, aged 62, who had previously botanized in Mexico, collected over 250 numbers in the hazardous Pongo de Manseriche of Peru. These exemplary exsiccatae were distributed to world's herbaria by Mrs. Floy Bracelin.

Cuatrecasas began his field explorations in Colombia. In the next decades these resulted in 25,000 collection numbers and numerous publications prepared at the Smithsonian and the Field Museum.

Fifth revised edition of Chamberlain's successful *Methods in Plant Histology* (ed. 1, 1901) published.

1933

Mycological Society of America founded.

McKelvey and Sax report cytologic evidence that chromosome morphology in *Yucca* and *Agave* invalidate the generally accepted phylogenetic position of the genera based on hypogyny/epigyny.

Morris Arboretum of 160 acres established at Chestnut Hill on the outskirts of Philadelphia as part of the University of Pennsylvania.

1934

University of Washington Arboretum of 267 acres founded at Seattle featuring *Rhododendrons*.

Herbarium of the University of Tennessee founded in 1888 by Lamson-Scribner destroyed by fire.

1935

Comprehensive *Manual of Grasses of the United States*, with line drawings, distribution maps, and full synonymy, published by Hitchcock, assisted by Agnes Chase.

Virologist Stanley reported the "Isolation of a crystalline protein possessing the properties of tobacco mosaic virus" in *Science*.

Blakeslee reported 250 genes had been found in *Zea mays;* about 40 in *Datura* with 200 others not yet localized.

Pollen Grains by Wodehouse summarized a decade of his investigation into their morphology as related to plant classification, paleontology, and medicine.

Cranbrook Institute of Science, Bloomfield Hills, Michigan, founded.

Flora of Panama with contributed accounts by several authors under editorship of Woodson and with supporting field work begun.

B. L. Robinson's death closed his 38 years as Curator of the Gray Herbarium and 29 years as Editor of *Rhodora*.

1936

Growth chamber developed at Boyce Thompson Institute.

1937

Avery, Burkholder, and others pursued quantitative studies on auxins.

Thimann suggested that differential effects of auxins in tissues may inhibit in one tissue and stimulate in another, and that tissues are marked by a series of overlapping optimal concentration curves.

1938

Fairchild's *The World was My Garden* published this year became a non-fiction best seller.

Fairchild Tropical Garden and Montgomery Palmetum of over 200 palms and 400 tropical trees dedicated at Coconut Grove, Florida.

1939

Ruben, Hassid, and Kamen first applied radioactive isotopic tracers to problems in photosynthesis.

Strybing Arboretum of 40 acres founded at Golden Gate Park, San Francisco.

Ames' *Economic Annuals and Human Cultures* published.

1940

Publication of *Chronica Botanica,* Verdoorn editor, began in Waltham, Mass., with Vol. 6 (1940-41) .

1941

Ruben, Randall, Kamen, and Hyde reported that the oxygen liberated in photosynthesis comes from the decomposition of water.

Using three mutant strains of the ascomycete pink bread mold (*Neurospora crassa*) Beadle and Tatum demonstrated that genes play a critical role in the specific biochemical processes of growth.

Commoner and Thimann conclude from iodoacetate concentrations in

coleoptile experiments that only a small fraction of respiration may be involved in growth.

1942

Loss of East Indian plantation quinine in World War II forced the U. S. Government through its Board of Economic Warfare (later the Foreign Economic Administration) to launch a field survey of wild *Cinchona* in northern South America.

Family Degeneriaceae, a primitive member of the Ranales, described by I. W. Bailey and A. C. Smith from Fiji collections.

1944

Living *Metasequoia* discovered in China three years after its recognition as a fossil genus.

First of 42 botanical expeditions to the sandstone Guayana Highland of northern South America directed by Maguire of the N. Y. Botanical Garden.

Sears proposed the word 'palynology.'

1945

"G. I.'s" issued Fighting Forces Series paperback edition of Merrill's *Plant Life of the Pacific World.*

1946

Stanley awarded Nobel Prize for his work on molecular nature of tobacco mosaic virus.

Svenson reexamined the Galapagos flora and found its reputed endemism exaggerated and that records from the Ecuadorian coast were conspecific.

1948

Duggar, after retirement at 71, discovered the first broad spectrum antibiotic, aureomycin (chlortetracycline), produced by *Streptomyces aureofaciens.*

Calvin and Benson found that the major intermediate compound in which photosynthetic carbon is fixed is phosphoglyceric acid.

University of Kentucky Herbarium destroyed by fire.

1949

Edgar Anderson opened a new field of investigation with *Introgressive Hybridization.*

Went first used fluorescent light on an extensive scale for plant experimentation at the Earhart Plant Research ("Phytotron") Laboratory, California Institute of Technology, Pasadena.

1950

Fernald published the 8th "Centennial" edition of Gray's *Manual.*

Stebbins published his important *Variation and Evolution in Plants.*

Model study on speciation in *Tragopogon* published by Marion Ownbey.

Journal *Ceiba* launched by Wilson Popenoe, L. O. Williams, and colleagues at the Escuela Agricola Panamericana, Tegucigalpa, Honduras.

1953

Structure of desoxyribonucleic acid molecule described by Watson and Crick.

1954

Arnon and co-workers discovered that excised chloroplasts can photosynthesize.

Goodspeed's *Genus Nicotiana,* a synthesis of cytogenetic studies, test garden plantings, herbarium and laboratory work, and of five field expeditions, published by Chronica Botanica.

1955

Total production of the five broad spectrum antibiotics amounted to over 450,000 pounds valued at approximately $195,000,000, of which tetracycline alone nearly equalled the other four drugs together.

1956

Botanical Society of America celebrated its golden anniversary with the award of "Certificate of Merit" to fifty botanists of all fields.

Mirov found that hybrids between *Pinus contorta* and *P. banksiana* show the inheritance of biochemical characters to be independent of morphological characters.

1957

17 individuals of *Pinus aristata* growing in the White Mountains of California found by tree-ring analysis to be over 4000 years old —one 4,600 years old—and therefore more ancient than *Sequoiodendron,* the oldest of which measures 3,212 years.

Rauwolfia, source of tranquilizing drug by action of reserpine, the object of a multiple study by anatomist-taxonomist Woodson, pharmacognocist Youngken, chemist Schlittler, and pharmacologist Schneider.

1958

Plant Science Catalog. Botany Subject Index published in fifteen volumes, making available in book form the 315,000 cards in the subject catalog of the U. S. Department of Agriculture—widely considered the most valuable bibliographic tool of its kind.

30,000 willow specimens assembled by C. R. Ball acquired by the U. S. National Arboretum.

1959

Hendricks and colleagues discovered phytochrome, a key enzyme in flowering.

Chromatography as a tool in the study of complex infraspecific hybrid relationships proved promising in study of *Baptisia* by Turner and Alston.

1960

"Climatron" opened at Missouri Botanical Garden. A geodesic dome 70 feet high, 175 feet in diameter, designed to display different microclimates.

Fourth and final volume of Abrams' monumental *Illustrated Flora of the Pacific States* published as completed by Mrs. Ferris and contributors.

1961

October. Hunt Botanical Library dedicated in Pittsburgh.

Calvin awarded Nobel Prize for his work in tracing sequence of reactions in photosynthesis.

Eames' long-awaited *Morphology of the Angiosperms* published.

1962

Gould published an *International Plant Index* employing an 80-column punched card designed eventually to give a reference number to each species.

1963

Sokal and Sneath's *Principles of Numerical Taxonomy,* primer for a new approach in systematics, published.

Plant hormone abscissic acid isolated and synthesized independently in the U. S. and England.

1966

Pollen wall ultrastructure as another source of data for assessing the taxonomic position of uncertain genera reported as promising in the Compositae by Skvarla and Turner.

Flora North America Project established with center of coordination at Smithsonian Institution.

1967

Arctic lupine seeds from Yukon Territory known to be 10,000 years old germinated by Porsild. Proved to be identical with living species.

1968

Hultén's *Flora of Alaska and neighboring territories,* "undoubtedly the most attractive and profound treatment of its vascular plants" (Constance), published.

BPH or *Botanico-Periodicum-Huntianum* published by Hunt Botanical Library, listing 12,000 serials that contain plant science papers issued between 1646 and 1966.

Dodson found that some orchids have fragrances composed of 18-20 chemical compounds effective both as general and specific attractants for euglossine bees.

Gray Herbarium Index, published in ten volumes, lists about 250,000 plant names for the Western Hemisphere.

Ginkgo planted in the Pierce Arboretum in 1789 (now a part of the Longwood Gardens, Kennett Square, Pa.) is now 105 ft. high, and nearly 13 ft. in diameter at three feet above the ground.

If a scientist is one jump ahead of his fellows in
his thinking he is usually their acknowledged
leader; if he is two jumps ahead he is thought
to be eccentric and rather screw ball but sometimes receives belated recognition in his old age.
If he is three jumps ahead he is ignored, though
posterity may eventually get around to appre-
ciating his evidence as it did with Gregor
Mendel.

Edgar Anderson

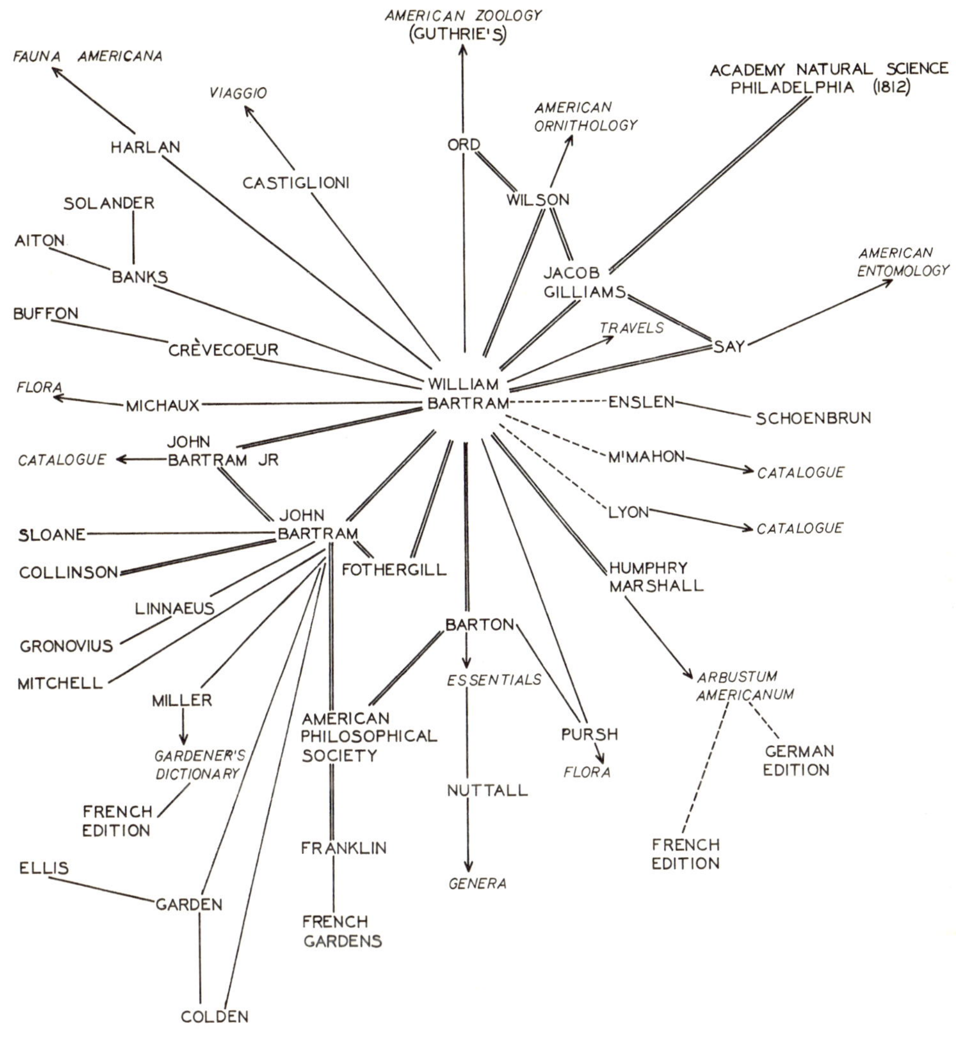

Fig.1. ROLE OF THE BARTRAMS IN NATURAL HISTORY

EARLY HISTORY

JOSEPH EWAN

(*Tulane University*)

Four hundred years have brought more changes in the vegetation of a continent than any other four centuries in the world's history. From a wilderness inhabited by a Brown Race living in widely positioned villages joined by trails which slipped along rivers and among the forests that blanketed the land, four centuries find urbs and suburbs peopled by two races, White and Black, crowded in cities strung on highways laced across the face of the continent. A primeval temperate forest of exceptional variety has by man's tools been reduced to woodlots and second growth. From barks, roots, fruits, seeds and saps to composition boards, synthetics, and confections. From a language of Pessemmins and Rakiock to *Diospyros virginiana* and *Liquidambar styraciflua.* Redwoods, ocotillo, bald cypress decimated, chestnut plagued, *Franklinia* extinct in the wild. Some endemic species of California lost even before their chromosome numbers were recorded.

In the same four centuries in which Europeans completed the conquest of the Americas, an eclectic and distinctive culture grew to maturity in the New World. America was invaded by observers, some merely curious travelers, some tireless naturalists whose gatherings laid the foundations, first for a workable classification of identified organisms followed then by the differentiation of the biological sciences into the multipartite specialties we know today. The passage of curiosities, reports, and hearsay across the Atlantic in accelerating pace through the centuries has been called by Theodore Hornberger "one of the most admirable traditions of the commonwealth of learning."

The development of botany in the United States is part of the intellectual history of the New World. From the often garbled commentary of the first European explorers punctuated by myth and prejudice, of potions and panaceas, the search for knowledge sharpened to specimens with labels, on-the-spot drawings, and comparisons with Old World forms—all important advances in the scientific process.

The history of botany in the United States may be arranged in eight epochs of varying duration characterized by persons or movements that have influenced its development (Table I). For example, Sloane, the master antiquarian, though he never visited this country, fostered the preservation of many collections that came into his possession. In

Table I the earliest *important* appearance of the person on the stage of history has been chosen as ushering in the period of interest in which he played a part. The last hundred years has been marked by such a gallery of figures that only a representation of the widening fields can be given. Table II, which summarizes the epochs of ornithology and geology, indicates that some of the same persons involved in their history are concerned with the development of botany, and this again stresses the wide, shared, 'naturalist' interests of our early scientists.

Although Santa Catalina Island was discovered off the southern California coast in October 1542 by Cabrillo, there was no botanist aboard the *San Salvador* or its consort, *La Victoria,* and it was 1847 before Gambel botanized on the island. From somewhere on the Atlantic coast seeds of *Thuja occidentalis* were taken to Europe. Belon mentions the tree in 1558 and by 1576 Clusius had reported its growing at Fontainebleau suggesting its origin to be New France. The description of Newfoundland by the French cartographer Thévet published in 1557 implies considerable contact between natives and Europeans. The botanist Ganong elucidated the Paris manuscripts of Thévet of whom it has been said that he is "most misread and the most condemned." The Spanish physician Monardes referred to Sassafras from "Florida" in 1569 by this, its French name, having received it about three years earlier from a Frenchman. Circumstances suggest that the specimen survived from the expeditions of Ribaut and Laudonniere (1562-66) to the "Riviere de May" (possibly the Altamaha River of Georgia rather than the St. Johns River) of northern "Florida" and to South Carolina.

Some other of the thirty plants known to Europe before 1600 (Table III) were from this expedition, and are probably among the Southeastern plants cited in Cornut's *Canadensium plantarum . . . historia* (1635), for example, Black locust (*Robinia pseudoacacia*). Although Cornut did not visit America and the greater number of his plants are included in an appendix, and are from the vicinity of Paris, he gave the first descriptions for about thirty mostly northeastern species with line drawings that are unmistakable and pleasing. Linnaeus frequently cited Cornut. No study has yet been made of the herbarium of twenty-three volumes of plants from Nouvelle France by the French apothecary Joachim Burser.

Certain plants were noticed because of their possible economic uses, for example, Maize and pumpkins; for their scent, as Pearly everlasting (*Anaphalis margaritacea*) ; their association, as with Columbine or Arbor-vitae; or because they were "of a shape so strange and monstrous" as Banister said of the Pitcher plant, or "fantastical" as Josselyn said. Gerard tells us that the southern Pitcher plant (*Sarracenia flava*) had been sent from Paris to Clusius who had received it

TABLE ONE

Epoch	Year	Names / Institutions
THEVET – JOSSELYN EPOCH	1555	THEVET
	1585-1600	JOHN WHITE – HARIOT
	1600-1674	V. ROBIN – TRADESCANT – CORNUT – JOSSELYN
SLOANE EPOCH	1678-1692	BANISTER
	1693-1722	VERNON – KRIEG – HUGH JONES – LAWSON
	1722-1730	CATESBY
BARTRAM EPOCH	1730-1747	JOHN BARTRAM – CLAYTON – STELLER – COLDEN – MITCHELL
	1743-	AMERICAN PHILOSOPHICAL SOCIETY
	1748-1751	KALM
	1753	LINNAEUS
	1754	GARDEN
	1770	ELLIS
	1780	AMERICAN ACADEMY OF ARTS AND SCIENCES – HOHENHEIM
	1781	WANGENHEIM
	1782	JEFFERSON – CRÈVECOEUR
	1783	JOHN BARTRAM JR. – WILLIAM YOUNG JR.
	1785	MARSHALL – CUTLER
	1788	WALTER – FRASER
	1789	BANKS (AITON)
	1790	CASTIGLIONI
	1791	WILLIAM BARTRAM – SESSÉ AND MOCIÑO
BARTON EPOCH	1797	*MEDICAL REPOSITORY* – COLLIGNON AND LA MARTINIERE (LAPEROUSE)
	1798	MENZIES (VANCOUVER)
	1801-03	A. MICHAUX
	1803	B.S. BARTON
	1804	*PHILADELPHIA MEDICAL AND PHYSICAL JOURNAL* LEWIS AND CLARK
	1806	HOSACK – M'MAHON
	1810	F.A. MICHAUX
	1812	ACADEMY OF NATURAL SCIENCES OF PHILADELPHIA
	1813	MUHLENBERG
	1814	PURSH – BIGELOW – RICH
	1815	W.P.C. BARTON
EATON – NUTTALL EPOCH	1816	ELLIOTT – EATON
	1817	LYCEUM OF NATURAL HISTORY OF NEW YORK WASHINGTON BOTANICAL SOCIETY – RAFINESQUE
	1818	NUTTALL – PECK – *AMERICAN JOURNAL OF SCIENCE*
	1819	LOCKE
	1820	PRINCE
TORREY AND GRAY EPOCH	1821	CHAMISSO (KOTZEBUE)
	1824	TORREY
	1826	DARLINGTON
	1829	ALMIRA HART LINCOLN
	1830	DOUGLAS (HOOKER AND ARNOTT) – HAENKE (PRESL) BOSTON SOCIETY OF NATURAL HISTORY
	1833	BECK
	1834	BOTTA (DUHAUT CILLY)
	1836	ASA GRAY
	1837	UNIVERSITY OF MICHIGAN MUSEUM
	1838	TORREY AND GRAY
	1839	WILKES EXPLORING EXPEDITION – HARTWEG (BENTHAM)
	1840	SULLIVANT
	1841	JOHN DARBY
	1845	ALPHONSO WOOD
	1846	SMITHSONIAN INSTITUTION
	1847	CHARLES A. SPENCER
	1848	AMERICAN ASSOCIATION FOR ADVANCEMENT OF SCIENCE
	1849	KELLOGG
	1852	RIDDELL
	1853	PACIFIC RAILROAD SURVEYS – CALIFORNIA ACADEMY OF NATURAL SCIENCE
	1854	PICKERING
	1856	ACADEMY OF SCIENCE OF ST. LOUIS
	1860	CHAPMAN
	1861	ENGELMANN

GRADUATE LABORATORY EPOCH

1862	MORRILL ACT
1863	NATIONAL ACADEMY OF SCIENCES – PORCHER
1868	PARRY – U.S. NATIONAL HERBARIUM
1870	*BULLETIN TORREY BOTANICAL CLUB* – SAUNDERS
1872	GOODALE (HARVARD)
1874	FARLOW (HARVARD) – *BOTANICAL GAZETTE*
1875	ATWATER
1877	*NATURALIST'S DIRECTORY*
1878	ARNOLD ARBORETUM – BURBANK
1879	BESSEY (NEBRASKA)
1884	LESQUEREUX AND JAMES – SARGENT
1885	*JOURNAL OF MYCOLOGY*
1891	NEW YORK BOTANICAL GARDEN – CAMPBELL – DUDLEY
1892	ATKINSON (CORNELL) – MACFARLANE (PENNSYLVANIA)
1893	BLASCHKA
1896	COULTER (CHICAGO)
1898	OFFICE FOREIGN SEED AND PLANT INTRODUCTION OF U.S.D.A
1899	STURTEVANT – *RHODORA*
1900	YALE SCHOOL OF FORESTRY
1902	SECOND INTERNATIONAL CONFERENCE ON HYBRIDIZATION AND PLANT BREEDING
1905	*NORTH AMERICAN FLORA*
1906	BOTANICAL SOCIETY OF AMERICA
1907	TROPICAL SCHOOL OF BOTANY, GUATEMALA
1908	EAST – SHULL
1913	*JOURNAL OF AGRICULTURAL RESEARCH*

FOUNDATION OR GOVERNMENT SPONSORED EPOCH

1916	*GENETICS*
1917	CINCHONA SURVEY IN COLUMBIA BY 3 INSTITUTIONS
1920	PHOTOPERIOD STUDIES AT U.S.D.A – *ECOLOGY*
1924	BOYCE THOMPSON INSTITUTE FOR PLANT RESEARCH
1925	JEPSON – NICOTIANA INVESTIGATIONS AT BERKELEY *TROPICAL WOODS*
1926	FOURTH INTERNATIONAL BOTANICAL CONGRESS, CORNELL *PLANT PHYSIOLOGY*
1932	*BRITTONIA* – *HARVARD BOTANICAL MUSEUM LEAFLETS*
1935	*CHRONICA BOTANICA*
1938	MONTGOMERY PALMETUM
1942	BOARD ECONOMIC WARFARE CINCHONA EXPLORATION PROGRAM
1944	GUAYANA HIGHLANDS MULTIPLE SPONSORED EXPEDITIONS
1950	*CEIBA* OF ESCUELA AGRICOLA PANAMERICANA
CONTINUING	HISTOCHEMISTRY BY RESEARCH TEAMS – ANTIBIOTICS – MOLECULAR BIOLOGY OF VIRUSES PHOTOSYNTHESIS – NUMERICAL TAXONOMY INDUCED MUTATIONS

EPOCHS, WITH SOME NAMES, DATES, AND EVENTS IN THE HISTORY OF BOTANY IN THE GEOGRAPHIC LIMITS OF THE UNITED STATES. NAMES SEPARATED BY DASHES MERELY FALL IN THE SAME YEAR(S). FOR ORGANIZATIONS AND FOR PUBLICATIONS (SHOWN IN SLANTING LETTERS) THE FOUNDING YEAR IS GIVEN. A PERSON WHOSE NAME APPEARS IN ( ) WAS RESPONSIBLE FOR MAKING KNOWN THE DISCOVERIES OF THE PERSON (USUALLY COLLECTOR) WHOSE NAME IMMEDIATELY PRECEDES HIM.

TABLE TWO

Epoch	ELLIOTT COUES			G.P. MERRILL	
ARCHAIC EPOCH	PRE-1700				
PRE-LINNAEAN EPOCH	1700-1730	LAWSONIAN	PERIOD		
	1730-1748	CATESBIAN	"		
	1748-1758	EDWARDSIAN	"		
POST-LINNAEAN EPOCH	1758-1766	LINNAEAN	"		
	1766-1785	FORSTERIAN	"		
	1785-1791	PENNANTIAN	"	1785-1819	MACLUREAN ERA
	1791-1800	BARTRAMIAN	"		
WILSONIAN EPOCH	1800-1808	VIEILLOTIAN	"		
	1808-1824	WILSONIAN	"	1820-1829	EATONIAN ERA
AUDUBONIAN EPOCH	1824-1831	BONAPARTIAN	"	1830-1839	STATE SURVEYS
	1831-1832	SWAINSONIO-RICHARDSONIAN			
	1832-1834	NUTTALLIAN PERIOD			
	1834-1853	AUDUBONIAN	"	1840-1849	STATE SURVEYS
BAIRDIAN EPOCH	1853-1858	CASSINIAN	"	1850-1859	STATE SURVEYS
	1858-	BAIRDIAN			

Table III

*Plants introduced from the United States and Canada into Europe
before 1600*

Agave virginica "Silk grass"	Hariot, 1588
Anaphalis margaritacea "Cudweede of America"	Gerard, 1599
Apios americana "Openauk"	Hariot, 1588
Aquilegia canadensis "Columbine"	Lobel, 1581
Asclepias syriaca "Apocynum syriacum"	Clusius, 1576
Asclepias variegata "Wausauke"	John White, 1585-87
Canna indica "Indian Reede"	Gerard, 1599
Castanea dentata "Chestnut"	Hariot, 1588
Commelina virginica "Branched spiderwort"	Gerard, 1599
Cucurbita pepo "Cucurbita indica rotunda"	Dalechamps, 1587
Diospyros virginiana "Medlar"	Hariot, 1588
Helianthus annuus "Planta solis"	Hariot, 1588
Helianthus tuberosus "Flower of the sunne, many on the stalke"	Gerard, 1599
Juglans nigra "Black walnutt"	Hariot, 1588
Juniperus virginiana "Cedar"	Hariot, 1588
Liriodendron tulipifera "Rakiock"	Hariot, 1588
Morus rubra "Mulberrie"	Hariot, 1588
Nicotiana rustica "Uppówoc"	Hariot, 1588
Opuntia humifusa "Metaquesunnauk"	Hariot, 1588
Rhus typhina "Shoemake"	Hariot, 1588
Sabatia stellaris "Gentian"	John White, 1585-87
Sarracenia flava "Limonio congeneris altera"	Dalechamps, 1587
Sarracenia purpurea "Limonio congener"	Clusius, 1576
Sassafras albidum "Sassafras"	Monardes, 1569
Smilax bona-nox vel aff. "Tsinaw"	Hariot, 1588
Taxodium distichum "Cypres"	Hariot, 1588
Thuja occidentalis "Arbor vitae"	Belon, 1588
Tradescantia virginiana "Spiderwort without branches"	Gerard, 1599
Yucca gloriosa "Iucca"	Gerard, 1599
Zea mays "Pagatowr"	Hariot, 1588

from someone in Lisbon who had received it from a Spanish monk. Explorers seemed uncannily attracted to plants of restricted range: *Yucca gloriosa* in its true form grows on the offshore islands of Carolina, but after reaching Europe was passed from garden to garden until it became a popular plant in cultivation. Gerard who called it *Iucca* was growing it before 1599, and Hariot was probably his source. Hariot, correspondent of Kepler and mathematician who introduced Copernican concepts into England, returned from an expedition financed by Sir Walter Raleigh to Virginia in 1586, not only with *Nicotiana rustica,* but with tubers of *Helianthus tuberosus* and *Apios tuberosa,* and *Cucurbita pepo,* for which he had learned uses from the Indians. Sweet gum attracted his attention with its echinoid fruits; the Staghorn sumac, a source of dye; the Persimmon, which he called medlars; perhaps the Black walnut; and because of his nation's basic interest in silk culture, the native mulberry (*Morus rubra*). The first drawings by a visitor came from this trip. John White drew Wausauke (probably *Asclepias variegata*) and a Marsh pink (*Sabatia stellaris*). John Tradescant the younger took back from his several trips to the lower York River district in Virginia between about 1632 and 1654, to his father's garden on the Thames' side, seeds or roots of nearly one hundred kinds. "Beare's ears" (*Dodecatheon meadia*) and Spiderwort (*Tradescantia virginica*) were two of these.

William Wood and Josselyn gave principal attention to plants of New England with potential economic values. Tuckerman (1865) has written a scholarly commentary for Josselyn's two accounts (1672 and 1674) which are often whimsical and illustrated by crude drawings.

William Penn inventoried the Indian crops and the potentially useful native plants. He was moved by the beauty of the Pennsylvania settlements granted him in 1681 by Charles II. Penn urged that homes be surrounded by gardens and orchards that each "may be a green Country Towne, w[ch] will never be burnt and will always be wholesome."

The earliest attributable preserved plant specimens made in the United States were for Robert Morison and his *Historia* (Vol. 3) by the Rev. John Banister, M.A. They are with specimens sent to Bobart, now part of the Sherardian Herbarium at Oxford. After Morison's accidental death Banister sent specimens to Ray, Lister, Compton, and Doody. These eventually passed to Plukenet and became part of the collection of Sir Hans Sloane, and are now in the British Museum.

Banister went to Virginia in 1678 with the idea of writing its "Natural History" and toward that end accumulated an impressive natural history library, but the financial insecurity of an Anglican minister in Virginia then, and the demands upon his time, slowed his progress so that his "Natural History" was incomplete at his accidental death in 1692. His "Account of the Natives" which contains observations on agricultural practices of the Indians, was neatly plagiarized by Robert Beverley in his *History and Present State of Virginia* (1708).

During his eleven years at Magdalen College, Banister, student chorister, clerk and chaplain, had become well acquainted with American plants growing in the Oxford Physic Garden and their literature, for he appears to have taken with him to Virginia an herbarium of carefully labeled specimens leaving the catalogue at Oxford for Morison, and later Bobart and Sherard, to compare with specimens he sent back. Bobart usually redescribed the new species for Morison, *Historia* (Vol. 3, 1699), but credited Banister in his commentary. Most of Banister's 79 plant drawings were reproduced in Plukenet's *Phytographia* (1691-1705), and most of his approximately 340 species, with his descriptive Latin names for the new ones, by Ray in his *Historia,* (Vol. 2, 1688, vol. 3, 1704). Banister's species in the works of Morison, Ray, and Plukenet, aided Gronovius and Linnaeus in their understanding of Clayton's specimens from Virginia, and his specimens were almost always behind the citation of those authors in Gronovius, *Flora Virginica* (Pt. 1, 1739, Pt. 2, 1743). Banister's introductions were soon growing in Bishop Compton's garden at Fulham, in the Oxford Physic Garden, and later at Chelsea and Leiden. After a futile search for another minister-naturalist to carry on Banister's work, Hugh Jones was sent to Maryland to collect in 1696, and in 1698 Vernon and Krieg. Their specimens are mostly in the Sloane herbarium at the British Museum (Natural History).

Clayton, Clerk of Gloucester Co., had developed an interest in native plants soon after his arrival in Virginia sometime between 1715 and 1720. He maintained a plantation about twenty miles from Williamsburg. Catesby while living in Williamsburg almost certainly met Clayton, and after his return to England introduced him by correspondence to Gronovius in Holland to whom he forwarded Clayton's specimens sent to him in England. Linnaeus worked with Gronovius over Clayton's specimens and Gronovius shared specimens with him so that there are two series of Clayton's specimens today: Gronovius' own, acquired by Banks and now in the British Museum (Natural History), and those Linnaeus added to his herbarium which is now at the Linnean Society. Clayton, obviously pleased to have his specimens appreciated, sent "A Catalogue of Plants, Fruits, and Trees native to Virginia" to Gronovius. Gronovius, without seeking Clayton's permission and probably on the urging of Linnaeus who was never one for loitering in publication, added to Clayton's original text which gave local names and uses, his own or Linnaeus' descriptive name, and synonyms from Morison, Ray, and Plukenet, and so brought out *Flora Virginica*. Ironically Clayton had no part in the revised edition (1762). He sent a manuscript on Virginia plants to England hoping Collinson would implement its publication, and several botanists annotated it, but it was returned to Clayton and was destroyed by arson in 1787.

When Catesby drew the Mock bird and Pigeon of passage, the

Ruffed grouse, and others for his *Natural History of the Carolina, Florida,* and *the Bahamas* (1730-43) he usually included plant backgrounds which he said were drawn "while fresh and just gathered." Catesby came to Williamsburg in 1712 for a short visit, but stayed seven years aside from a short Jamaica tour. Samuel Dale and Thomas Fairchild grew seeds he sent in their gardens and sent reports to Sherard. In 1722 Catesby again sailed, this time to Carolina where he remained until 1725 and with financial support from, among others, Sherard, Sloane, and Dubois. His specimens are preserved at Oxford in the Sherardian Herbarium. Seeds from his Carolina stay were grown at Chelsea and at Hoxton where Catesby settled on his return. Pursh and Gray examined and annotated Catesby specimens at Oxford. With all the flurry of feathers over Catesby there has not been a correlated botanical study of his collections against his letters, scattered notes, and textual references in his *Natural History*. His posthumous *Hortus* has been generally dismissed as a publisher's sinecure. Completed before his death it dealt with 85 American trees and shrubs for English gardens, of which 63 were illustrated on seventeen plates, all based more or less closely on the *Natural History* drawings. The text of the *Hortus* was, however, amplified over the plant notes in the larger work, and included some gatherings from John Bartram from whom he had first received a letter in 1740. Linnaeus based twelve binomials in *Species plantarum* on Catesby.

While Catesby was engaged with his *Natural History,* two botanical activities were stirring in the Southeast. Both passed into history as withered leaves. Bolzius in the Moravian colony at Ebenezer near Savannah, Georgia, was one of seventy-eight Protestant refugee Salzburgers who settled in Oglethorpe's new colony. Georgia was established as an asylum for debtors and persecuted European Protestants; a trial garden for the growing of silk, wine grapes, and drug plants; and as a relief of population pressures in Mother England. From 1736, after Old Ebenezer had moved to nearby higher healthier ground, until 1774, the colony prospered with brief success in its silk culture, but the lack of appreciation by the employers of the necessary patience, training, and skill in their employees, and the unfavorably hot and sticky climate, led to its demise. Bolzius, who had authored a propaganda tract, "Ample Directions for the Colony of Salzburg Emigrants in America" (1736) which focused on economic plants, rice, tobacco, mulberries, potatoes, maize, grapes, and the like, wrote what amounts to the first local flora for Georgia wherein eighty economic plants, native and introduced, are noticed. It was published anonymously in Hamburg (1756). Bolzius frequently cited Catesby and Gronovius, *Flora Virginica*. Six oaks, five grasses, eleven *Rosaceae,* and six legumes are among them. Nine new species appear under vernacular names.

Houstoun, who had botanized in the West Indies, Vera Cruz, and

the Spanish Main on behalf of Sir Hans Sloane and the Chelsea Physic Garden, was retained as physician and naturalist by Governor Oglethorpe and the South Sea Company with the hope that he would ship plants and seeds from tropical America to West Indian ports, and to Charleston and the "Colony of Georgia in America." His untimely death in Jamaica in 1733 brought such plans to an end though some West Indian plants did reach Savannah plantations.

The first native American botanist, John Bartram, was both a self-taught farmer and an original member of the American Philosophical Society. His farm near Philadelphia was more than a nursery, yet strictly less than a botanical garden. It was a meeting place of naturalists, a mecca for travelers, and because his son William never married but continued to live on in his father's stone house, Bartram's garden blosomed, though with declining vigor, until William's death in 1823. John Bartram's association with Franklin in Philadelphia, and his Quaker comradeship through correspondence with Collinson abroad, enhanced the gathering of plant lore. These ties brought Colden who lived on the Hudson River, Garden of Charleston, Lord Petre, Philip Miller, and many others, into Bartram's circle. His specimens, often with quaint notes on their medicinal uses and folk names, are mostly in the Sloane Herbarium but a small overlooked series forms a part of Lord Petre's *Hortus siccus stirpium Americanum* at the Sutro Library, San Francisco.

The Bartram Epoch was characterized by the revival of a lively traffic in seeds and plants from America to England (and on to the Continent) which continued until John's death in 1777. Darlington fortunately personalized these decades for us with the *Memorials of Bartram and Marshall* (1849). From Collinson's letters to Bartram some understanding of the trials of trafficking in perishable seeds may be gained. "Prithee, go at a proper season to the nearest place, and load a pair of panniers or baskets, with young plants, and set some in thy garden to take root, and send half a dozen at a time." "Another time if thee sends any growing plants, a great many may be packed close together in a case two feet square, or two feet wide, and three long." Casks, barrels, or boxes, if not stowed under a berth, were set on the unprotected deck, at mercy of winds and waves, and most devastatingly of the "mischievous and unruly vermin," the "young callow rats." Then there were the "prying, knowing people" who stole the rarer sorts between ship and carriage. "Somehow the picture of a rolling, tossing, small ship, the long days and weeks of transatlantic trips, and these boxes of roots, hastily thrust under the berth to keep them from cattish curiosity, brings out the true character of the eighteenth-century sea captains," wrote Sarah Stetson (1949). "They were not only hardy, salty tyrants, they were the most forebearing mortals who ever walked a quarter deck. Cats, planks, sharp-cornered

boxes, heavy seas—anyone who could endure that combination in a tiny ship's cabin was a true martyr to science."

A literature had accumulated on this transport of living plants. Among these were Petiver's "Brief Directions for the Esie Making, and Preserving Collections" of about 1700, Samuel Pullein's *Observations towards a method of preserving seeds of plants in a state of vegetation during long voyages* (London, 1760) a copy of which has not been located, but this would be a decade before Ellis's *Directions for bringing over seeds* and *plants* designed to "look out for a safe way of conduct."

The apogee of the Bartram Epoch came with the travels of John and William Bartram into southeastern United States under the patronage of Collinson and Fothergill. By a cruel twist of fate the specimens, descriptions, and drawings of the many discoveries they made were given manuscript names and set aside in the cabinets of English botanists without publication while Britishers were engrossed with the plants of Cook's voyages to the exciting South Pacific. Other botanists revisited Bartram localities, some of them after calling at the Bartram garden to learn particulars, and described Bartram discoveries in print under their own names. While Fothergill and his circle brought novelties to flower, Banks and Solander studied and judged the characters of twenty-three "firsts," some of them drawn by William. Genera now known as *Balduina, Befaria, Chapmannia, Chaptalia, Elliottia, Franklinia, Glottidium, Macranthera, Mayaca, Pinckneya,* and *Polypteris,* were all Bartram discoveries. Star anise was collected in northern Florida in 1766 by the Bartrams nine years before John Ellis announced its discovery and named it *Illicium floridanum.* The Bartrams' broadside "Catalogue" [1783] appeared with *nomina nuda* that were to be given descriptions in the *Travels,* publication of which was delayed until 1791. In the interval Marshall's *Arbustum Americanum* (1785), Walter's *Flora Caroliniana* (1788), Aiton's *Hortus Kewensis* (1789), Lamarck's descriptions in the *Encyclopédie Méthodique,* and L'Heretier's *Stirpes Novae,* had all publicized Bartram's discoveries from the Carolinas, Georgia, and Florida. From his father's death in 1777 until his own death in 1823, William Bartram (Figure One) was the center of influence in American natural history, not only of botany, but also especially of ornithology.

John Mitchell, Physician turned planter who lived on the Rappahannock River between 1735 and 1746, is said to have taken his M.D. degree at Leiden where he was most certainly influenced by Boerhaave. His papers were on the "principles" of botany, on the opossum, yellow fever, the "causes of the different colours of People in different climates," and on potash and the "force of electrical cohesion" —the last in the Royal Society's *Philosophical Transactions.* Although he corresponded with Sloane there are no discernible Mitchell speci-

mens in the Sloane Herbarium since pirates seized some of his collections, but there are Virginia specimens of his in the Sherardian and Dillenian herbaria at Oxford. Mitchell's correspondents included Franklin, Linnaeus, Colden, Bartram, and Clayton, although letters between Mitchell and Clayton have not been found. His map of the "British and French Dominions in North America" (1755) is certainly his best known accomplishment. Biographers disagree on his birthplace (England, Scotland, or Virginia), the length of his stay in Virginia (between 11 and 47 years), when he died (sometime between 1768 and 1772) and whether he was the author of the anonymously published *American Husbandry, containing an account of the soils, climate, production and agriculture of the British Colonies in North America and the West Indies* (2 vols., 1775). The multiplicity of John Mitchells complicates the problems. His place in botany is underfoot: Linnaeus named the demure Partridgeberry (*Mitchella repens*) for him.

Peter Kalm arrived in Philadelphia aboard the *Mary Gally* on September 15, 1748, traveled into Delaware, New Jersey, and as far north as Quebec, and returned to Sweden two and a half years later with specimens for his beloved mentor, Linnaeus. Kalm's *Travels*, resulting from his careful notes, continues to amuse and inform us of colonial America, and he has been studied by Juel in Sweden, Daydon-Jackson and Savage in England, and Larsen and Benson in this country. His botanical collections, extensively cited by Linnaeus in *Species plantarum*, represent types for many of our northeastern United States species. The source of some is hidden behind the phrase "in Pennsylvania paludosis" which may mean the margins of the Pine Barrens in Salem County, New Jersey. His visit with Bartram late in November 1749 was especially cordial, and references to botanical matters recur in many letters.

Like some of Mitchell's, Colden's collections destined for Europe were seized by pirates. In the course of his work as surveyor-general for New York State he had an exceptional opportunity to discover new plants in the uplands and mountains. He wrote to Collinson and Gronovius on botanical topics, and Linnaeus published his "Plantae Coldenghamiae in Provincia Noveboracensis sponte crescentes," in *Acta* of the Royal Academy of Uppsala, in two parts (1749, 1751), of the plants around his home in Orange County, New York. Gray published certain Colden letters (1843). Medical botany interested Colden, such as the Indian use of *Lobelia* as an antisyphilitic. He encouraged his daughter Jane to describe and draw the local plants: in fact her crude sketches may have challenged fourteen year old William Bartram, while on a visit to Coldengham with his father, to strive for improvement of his own drawings. Kalm and Garden were also visitors to the Coldens.

The Charleston physician-naturalist Alexander Garden is honored

ironically by an exotic plant. His importance in botany rests more on his support of Bartram and Colden in America, and his contacts with Linnaeus and Ellis abroad, than on his Carolina plant discoveries. His zoological interests, especially his Carolina fish collections were of more critical importance.

On the opposite side of the continent in 1741, a Danish navigator, Vitus Bering, sailed into the sea which bears his name aboard the Russian packet-boat *St. Peter* accompanied by the naturalist Steller. They first encountered the edge of Alaska at Kayak Island east of Seward. Steller was ashore only six hours, hustling to observe, collect, and record *naturalia*. He compiled a catalog of plants on the spot listing the characteristic coastal species under Latin phrase names, many not identified until two hundred years later by Stejneger. Steller described the Salmonberry and tried to take living plants back to St. Petersburg. It was formally named *Rubus spectabilis* seventy years later. About thirty of Steller's Kamtchatka specimens are in the Linnaean Herbarium but his first collections made in Alaska were lost.

The first serious study of New England plants was Cutler's "An Account of some of the Vegetable productions, naturally growing in this part of America, botanically arranged," published in the first volume of the *Memoirs of the American Academy of Arts and Sciences* (1785). No new names were proposed, hence the paper is not well known. Indeed, the names are *American* vernacular, the author adding that "from want of botanical knowledge, the grossest mistakes have been made in the application of *English* names of *European* plants, to those of *America*." Cutler offers no binomials even though the arrangement is Linnaean Sexual System. Withering and Cullen are cited, and such practical matters are emphasized as ridding fields of crows by boiling "Indian corn" in a strong decoction of White hellebore roots and "strewing it on the ground where they resort." Cutler made his first ascent of Mount Washington in 1784, but the collection was ruined on the descent. In 1787 he visited Ezra Stiles at New Haven and showed Stiles, his wife, and "the young ladies" his botanical "apparatus" and books "with which they were all highly pleased, having never seen any thing of the kind before." He demonstrated how by counting stamens and pistils "separating and exhibiting the parts" the specimens might be classified. The company was "highly amused," and he had to "explain technical terms, and construe crabbed Linnaean Latin for an hour on a stretch." Professor Peck of Harvard joined Reverend Cutler and party in 1804 on his second ascent of Mount Washington. Peck took a few specimens which he later showed Pursh; among them was *Geum peckii*.

John Fraser took more American plants to English gardens than any other person, according to Roscoe of Liverpool. Fraser was an enterprising nurseryman of Sloane Square, Chelsea, who made at least

seven trips across the Atlantic, and traveled over 4,000 miles in the seaboard states. In 1806 he opened his growing garden at Charleston, where he (or his son) were host in 1809 to another plant hunter John Lyon, and they compared fortunes in finding magnolias. Lyon recorded in his journal that on June 3, 1803, he found six or eight full grown trees of *Franklinia* growing near Fort Barrington, Georgia: he was the last person known to have seen them in the wild. These Britishers took back their prizes for European gardens, but had slight effect on American tastes in horticulture. Minton Collins of Richmond, Virginia, advertised "Garden and Grass seeds with a choice collection of flower roots, & seeds just imported" in 1793. Persian iris, "Red, White, & Blue Hyacinths," "Large Nerstertion" from Peru and "Gerenium" from South Africa were among "the best of all sorts" offered through "a very particular friend in London," so Americans were reaching for exotics while Englishmen bought Appalachian novelties!

It was Fraser who put Walter's *Flora Caroliniana* (1788) into print, who featured "Agrostis Cornucopiae," and whose sons announced Nuttall's prairie novelties in a broadside (1813). David Landreth, successful Philadelphia nurseryman, featured Lyon's natives, Magnolias, Stewartias, etc., for American gardens though he gave prominence to buxom cabbages over ornamentals.

The Barton Epoch opened in 1789 with Benjamin Smith Barton teaching Materia Medica and natural history at University of Pennsylvania. Adam Kuhn had held that position since 1768 but, though a student of Linnaeus, he gave scant impetus to American botany. Waterhouse was teaching at Harvard a year before Barton began in Philadelphia, and MacLean took a Princeton appointment in 1796, but gave more emphasis to chemistry than botany. Mitchill was appointed professor of botany at Columbia in 1792, and Drowne at Brown University in 1811. One of Barton's students, Eli Ives, spurred medical botany at Yale, and in 1814 invited Pursh to superintend a botanic garden, but Pursh declined in favor of Lord Selkirk's Red River Valley exploration.

What is generally mentioned as the first botanical textbook in the United States is Barton's *Elements of Botany* (1803), a philosophic rambling discourse filled with quotations gleaned from his extensive library, certainly the best in the country for natural history. His *Fragments of the Natural History of Pennsylvania,* a butt of his critics' scorn, was one of several more or less unfinished writings. More significant was his effort to launch a "Flora of North America." To that end he supported first Pursh and later Nuttall for field excursions north to Canada, west to the Great Lakes, and beyond. Barton's herbarium is evidence of his ambition to know the native plants. Though he failed to realize his grand plan, and though he was disliked by his fellow naturalists for his parsimony, flirted with deceit in some

of his transactions and hampered the Lewis and Clark enterprise, Barton taught Baldwin, Darlington, Ives, Horsfield, and many less well remembered students. He played decisive roles in the lives of William Bartram, Pursh, and Nuttall.

The American Revolution changed the direction of botany from dominantly English contacts to internal, and for a time, Franco-American. French naturalists such as André Michaux and his son François, Palisot de Beauvois, Bosc, Delile, Milbert, Claude Robin, Lakanal, and others, journeyed far, botanized, visited with Americans, and reported, to leave an impressive literature. Best known was André Michaux who came in 1785 on behalf of his government to search for timber, carpentry woods, medicinal and crop plants. His son François-André, then fifteen, accompanied him, and a journeyman gardener, Saunier by name. Three days after landing Michaux was collecting plants in New Jersey, and within two months had sent twelve boxes of acorns and seedlings to France. Michaux traveled from the southern shores of Hudson Bay and Lake Mistassini in Quebec to Charleston, where he opened a growing garden, and northern Florida, into the mountains of Kentucky and Tennessee, and to the plains of southern Illinois. He had opened a growing garden, known locally as the "Frenchman's garden," six miles from New York in present Hudson County, New Jersey, and put Saunier in charge. It operated on behalf of the French government until 1792, and was visited over the years by Cutler, Pursh, Bradbury, David Douglas, and other botanists. Of all Michaux's introductions *Rhododendron catawbiense* was the garden spectacle. One of his discoveries puzzled Michaux and he left the specimen, labelled "Hautes montagnes de Carolinie," in his herbarium unnamed. Gray examined it in 1839 in Paris, sought to find the prize himself, but it was 1877 before Hyams rediscovered *Shortia,* as it was named, on the Catawba River of North Carolina, and nine years later by Sargent at the headwaters of the Keowee River. It was partly Michaux's *Flora Boreali-Americana* (1803) that prodded Barton to begin a "North Ameircan Flora." This in turn led Pursh under Lambert's spur to publish his two-volume *Flora,* ten years later in London. François-André Michaux followed his father's work on the oaks published in 1801 with *Histoire des arbres* (3 vols., 1810) later published in English through the encouragement of Maclure, and much later considerably enlarged by Nuttall. Before Michaux returned to France he gave the American Philosophical Society a bequest, operative today, to further studies in American forest trees.

As Professor Barton fostered botany in Philadelphia, so did Doctor Hosack in New York. Both had been medical students at Edinburgh, both taught medical students after their return, both saw the need of a flora of North America, and to that end in turn each supported Pursh, and both devoted time and personal resources to a botanic

garden—Barton to one connected with the Pennsylvania Hospital at Philadelphia, and Hosack to the Elgin Botanic Garden. Both were active in local scientific societies, founded journals devoted to medicinal botany, each collected a notable private library, and each left an impressive list of publications. Yet in personalities they were very unlike. Hosack, "an eminently clubbable man," took pleasure in befriending visitors, making introductions, exchanging seeds, giving away his books, and in sharing his madeira and his fortunes. Hosack's *Hortus Elginensis* (1806; ed. 2, 1811) listed native and exotic plants cultivated in his garden, then on the outskirts of New York City at the site of the present Rockefeller Center. His biographer Christine Robbins has related the chain of circumstances that connected Eaton, Torrey, and Gray with Hosack's "pioneering venture, the first public botanic garden in the U. S." Where the Bartrams had grown a wide variety of transplanted natives of both useful and ornamental value, Hosack gave attention to species of interest to Materia Medica although he also grew *Aucuba japonica* and *Gleditsia sinensis* for *public* demonstration.

Doctors Hosack and Mitchill instructed a law student in botany, chemistry, and natural philosophy, who was to turn to botany, and thus "popular instruction in Natural History in this country began with Amos Eaton." Eaton, in turn, profoundly influenced the son of the prison's fiscal agent while imprisoned because of a controversial legal involvement as a land agent. He taught the lad, John Torrey, the rudiments of flower structure. Eaton wrote his wife from prison that he had "contrived a new method of arrangement, by which I can exhibit all the known species of plants (about forty thousand) in one small duodecimo volume,"—this employed the Sexual System—and so after his release, and as a student in Professor Silliman's class at Yale in chemistry, and Ives's in botany, Eaton improved his time by translating Richard's botanical dictionary. This he published anonymously in New Haven in 1816, and again in 1817. His *Manual of Botany* went through eight editions from 1817 to 1840, all the while pledged to the Sexual System despite the rising Natural System. Eaton's *Manual* was *the* field reference book for every botany student in the higher schools and academies of the time, the forerunner of Gray's *Manual*. Eaton influenced many students at Albany: Rafinesque, Riddell, and a remarkable woman, Mrs. Lincoln, whose *Familiar Lectures in Botany* (1829) became just that to a generation of ladies enrolled in "female seminaries." Over a period of forty years 275,000 copies of her *Botany* were sold; in the tenth edition (1840) she commented that "obstacles [have been removed] which formerly impeded the progress of botanical information, in schools, and among our own sex. We have seen that even children may become botanists, and lay aside their toys to divert themselves by distinguishing the organs of plants and tracing out their classification." She noticed that "Animals, though affording the most striking marks of designing wisdom, cannot be dissected and examined without painful emotions. But the vegetable world offers a boundless field of inquiry, which may be explored with the most pure and delightful emotions." Botanical language included some unmentionables, hence germ was suggested for "ovary" and placenta was avoided by circumlocution. Linnaeus, her mentor, had used the Latin *torus,* for marriage bed on which stamens and pistils, the *genitalia,* disport themselves, but these terms are absent from both Mrs. Lincoln's and Eaton's writings.

Nuttall was 22 when he arrived in Philadelphia and soon met Professor Barton, who instructed him in points of botany, urged him to make use of his library, and sent him forth to collect, at eight dollars a month, without rights of publication, but with a share in his collections. His biographer Jeannette Graustein says that "although Barton obviously considered him expendable, Nuttall miraculously survived all dangers to become a productive scholar." In 1844 Gray aptly summed up Nuttall's accomplishments: "no botanist has visited so large a portion of the United States, or made such an amount of observations in field and forest. Probably few naturalists have ever excelled him in aptitude for such observations, in quickness of eye, tact, in discrimination and tenacity of memory." His *Genera of North American Plants* (1818), designed to supplement Pursh's *Flora,* was a remarkable summary of his own explorations described with the benefit of the collections in the Philadelphia Academy where he worked night and day on their characters. Dr. Baldwin, a student of Barton, never enjoyed good health and died at the age of 41 in Missouri as a member of Longs Expedition. Baldwin's treatment of a number of genera especially in the *Cyperaceae* showed penetrating observation, understanding, and diagnosis. From Baldwin's letters, published by Darlington (1843), may be gained a candid perspective on the first two decades of our nineteenth century botany.

During all these years Philadelphia continued to be the center of intellectual activity, seat of the important University of Pennsylvania with its medical school, site of the most important Academy with a growing library and herbarium, and of many other influential societies, the publishing center of the nation, and a leader in commerce and industry. By Jefferson's death a change had set in which brought a shift in the population and scientific activities of the nation. The population of the country had increased ten fold in Jefferson's life time, 1743 to 1826, and the scientific stage moved from Philadelphia to New York and Cambridge, and later to Washington. The short-lived "backwoods Utopia" at New Harmony was one of the early migrations that took scientific talent to the west of the Allegheny Mountains. To Lexington, Kentucky, went lonely Rafinesque, ponder-

ing the changes that take place in species from generation to generation, arriving at Transylvania University during the administration of
its President Holley who evidently appreciated the "little, dark,
foreign-looking professor." Rafinesque's years at Transylvania probably contributed most to his reputation. He was succeeded at Lexington in 1825 by the retiring but competent Dr. Short who remained
thirteen years, teaching, botanizing, and corresponding with botanists
far and near.

In 1833 Professor Eaton declared that "since Dr. Faustus first exhibited his printed bibles in 1463, no book has, probably, excited
such consternation and dismay, as Dr. Torrey's edition of Lindley's
Introduction to the Natural System of Botany." Eaton did not believe
that Lindley, Hooker, Loudon, or deCandolle, who all had fostered the
Natural System, had published any improvement useful to the student
of North American botany. The first lectures on the Natural System in
the United States were presented in Philadelphia in 1815 by Correa
da Serra who published an adaptation of Muhlenberg's *Catalogus*
"for the gentlemen who attended the course." Torrey promulgated
the Sexual System from 1819 with the appearance of his *Catalogue*
until 1826 and his *Compendium of the Flora of the Northern and
Middle States*. But in that same year he changed his view and adopted
the Natural System in his report on the plants collected by James in
the Rocky Mountains.

The *Flora of North America* was planned, fostered, and stands today
—it is being reprinted after 130 years—as Torrey's monument, which
Gray helped to raise. When Torrey journeyed to Glasgow in 1833 as
an agent of New York University he discussed the plan for the *Flora*
and won quiet support of Hooker in its persuance. Two years later he
commented that the *Flora* was "an extensive work which I cannot
expect to finish in several years. I shall not publish prematurely."
Gray worked hand in glove with Torrey, as Bentham and Hooker
worked across the Atlantic. Gray's trip to Europe in 1838 was the first
trip of an American botanist to visit European museums expressly to
study type specimens essential to the writing of a flora. That he worked
rapidly is evident from the annotations he left on specimens at Oxford,
London, and Paris. Until 1842 when Gray went to Harvard and
Torrey to Princeton the team worked closely on the *Flora*. Six of seven
parts were published before the team separated, but Torrey clung to
the hope that the *Flora* would be finished. As late as 1847 he called it
a "*chronic* call," but the many other botanical demands brought its
abandonment.

Though Torrey lacked the "philosophical breadth of his pupil and
protégé, Gray," remarked Geiser, "he was an admirable taxonomist,
and left an impress on all subsequent taxonomic work." Gray was an
important refutation of what Smallwood (1941) believed had been the

prevailing attitude, that Americans failed to recognize that observation, classification, and verification—all three—must precede deduction. "The naturalist fitted his explanation to the current philosophy of his period, while the scientist lays the foundation for the philosophy of the next generation." Gray, with his allegiance to Darwinian postulates, his attention to Sino-American floristic patterns, and global principles, gave significance to what in themselves were insignificant botanical data.

The United States Exploring Expedition, generally called after its commander the Wilkes Expedition, was one of the most consuming enterprises of the mid-century: consuming of monies and manpower. Bartlett (1940) rightly evaluated its greatest value as the enormous coordinated effort of a few scientists that "enabled America to take a place with the leading European nations as a partner in the development of world science." The Reports born of the Expedition "are a monument in the record of American science," though inadequate and ineffective distribution greatly reduced their influence. The birth of the Reports is a harrowing tale; let two quotations from letters of the times speak. Brackenridge, who wrote the fern account, wrote to Torrey in 1851, "I am heartily sick (after all my labour with yours combined) in contemplating the present state of things, and what may be the end of this crooked affair." And Wilkes wrote Senator Tappan of Ohio, three months later, "I need not tell you we sometimes get into *tight* places, especially in having to deal with the whims and caprices of the gentlemen of science . . . I am sure you would . . . have a hearty laugh over the follies and conceits of those who confine themselves to one branch of science, and consequently believe it of more importance than all other subjects and views that can be entertained by others." The *national* character of the project is emphasized by the original plan that all authors of the Reports be American, "a sort of scientific declaration of independence." Gray succeeded in getting the restriction lifted for Harvey's essay on the algae, and Berkeley's on the fungi, but an American had to stand in as joint-author for each. The subtitle of one official volume read, not from an error in proof-reading but through patriotism, "Bailey and Harvey." The appointment of Rich as botanist to the Expedition was unfortunate. Gray wrote to Hooker after the return of the ships that "the botanist who accompanied the expedition is no doubt perfectly incompetent to the task, so greatly so that probably he has but a remote idea how incompetent he is." Nevertheless, the genus *Richella* and eighteen species were named for him by botanists! One of the most important results of the Expedition was that American botanists took a geographically unrestricted view of botany. From the time of the Wilkes Expedition American herbaria have grown to international importance.

Another restless activity involving both the Government and Gray's

botanical comrades was the procession of explorations and surveys "to ascertain the most practicable and economical railroad route from the Mississippi River to the Pacific Ocean." The "Pacific Railroad Reports" based on these explorations included botanical papers by Torrey, Gray, Engelmann, and introduced to the botanical public Parry, J. M. Bigelow, and others who would play out their parts during the next decades. The Reports, bibliographically, are the "wrecks of matter and the crush of worlds." Dates on the title page of the volume, of the initial leaf of separate reports within the volume, and dates scattered through the text, often disagree. An octavo unillustrated first edition consisting of two volumes only is almost unknown; the quarto second edition of volumes 2, 4, 5, 6, 7, and 12 (part 2) is illustrated with fine plates. Volume two was issued in two formats; the text differs in each. Ivan Johnston (1943) provided a bibliographic analysis of these Reports. As with Wilkes Expedition reports, the pains of publication smarted for years. Torrey lamented that the authorities were prevailing since Lt. Whipple insisted "this is the only way, or the Natural History may be thrown out altogether," and Torrey lamented that he found so many typographical errors.

Yet these joint military-civilian enterprises were important for botany in the United States. Historian Dupree remarks that remote New Mexico was "brought into the purview of international science within a few years after 1846" because civilian collectors accompanied the troops called up with the outbreak of the Mexican War. Fendler collected many undescribed plants along the Santa Fe Trail with the army; Wright entered El Paso as a collector with a military wagon train, and so on. Dr. Coues and other United States Army surgeons also actively participated in collecting *naturalia,* and hundreds of plants of the Far West bear commemorative names dating from this government exploring and surveying period.

It is told of Henry Shaw that when he planned a botanic garden, now that the heart of the nation had come of age, and sought advice of Hooker at Kew for the name of a botanist to guide his undertaking, he was told that the man lived at Fifth and Walnut in St. Louis and that his name was George Engelmann. Fortunately Engelmann's advice to Shaw was to develop his garden not only as a pleasure park but as a botanic garden of scientific distinction. A busy physician, Dr. Engelmann's botanical accomplishments were his avocation. Sargent's verdict that "no other American botanist has ever worked out harder problems or elucidated so many different groups of plants," still stands. Cacti, his first study, based on collections brought to him from the Pacific Railroad Surveys, then the oaks, dodders, euphorbs, rushes, conifers, mistletoes, the Century plant and its allies, yuccas—he noticed the obligate pollination relationships of the *Pronuba* moth in Shaw's garden—and the grapes, all were systematically studied and reported

on. He wrote Gray that he had purchased in 1857 at Leipzig the herbarium of Bernhardi of about 40,000 species. He persuaded Shaw to substitute "Missouri Botanical Garden" for the Latin on the entablature at the garden gate. Gray hoped Engelmann would accept the directorship of the garden "this duty must devolve upon you, and when it does, with a decent salary, you could reside up there, throw physic to the dogs, or take a share in consultations, and have time to do yourself justice in botany." But Engelmann found Shaw "tough as any Scotchman, and I fear for the present he won't buy any more books . . . a man who has no real scientific zeal nor knowledge who must be got to do things by diplomacy, I can not do much with." Gray, enthusiastic about the "right development of the Mississippian Kew," was consulted about details down to the mounting paper for the herbarium. Hooker told Shaw that Gray was prouder of Shaw's having consulted him on plans for his garden "than of any of the many services which he had rendered to American botany."

Never again will a single person dominate American botany as Asa Gray held dominion for fifty years. A man of sound scholarship, broad culture, of compelling personality, yet was he jealous of his pre-eminence. Alphonso Wood, without the talents nor the advantages of Gray, competed successfully in the textbook field. The first edition of Wood's *Class Book of Botany* (1845) promptly sold out; the second edition and subsequent printings from stereotyped plates held their popularity; all the while the author was traveling and learning new floras. He made a major revision in 1861 and others followed throughout his lifetime. Altogether between 800,000 and one million copies of his botanies were sold. During 1865-66 Professor Wood visited the Pacific Coast from San Diego to Cowlitz River, Washington, and incorporated in his later editions descriptions of these new materials with abbreviations to indicate distribution. Bolander, author of a catalogue of limited popularity on the plants of the San Francisco area (1870), and Superintendant of Education for California, wrote enviously to Gray in 1866, "What is to be done with Professor Wood?" Rattan, another of Gray's retainers, a teacher in San Francisco and the normal school in San Jose, published *A Popular California Flora* (1879) which, with frequent printings and minor revisions, held the market for two decades.

Whereas Professor Wood made independent decisions about the plants he found on his brief trips into the Southeast as far as Apalachicola, Dr. Chapman, a resident of Florida for over fifty years, sent his collections to Torrey and Gray, closely patterned his *Flora* after their work, and insisted he had no interest in proposing new species under his own name. He did inadvertently make some new combinations in the course of revising his *Flora*. Chapman was an Amherst graduate who became principal of the Academy at Washing-

ton, Georgia, where he studied medicine. He moved to Florida from where by 1838 he sent specimens to Torrey and Gray. Gray called him an "accurate and indefatigable botanist" and named a genus *Chapmannia*. Chapman asked forbearance for not discovering a parade of new species because he found Seminoles in the southern swamps! By 1845, nevertheless, he had sent 713 species of Florida plants to the British Museum. From 1847 until his death in 1899 he lived in Apalachicola, which for botany became synonymous with Chapman as Oquawka is synonymous with Patterson. His *Flora of the Southern United States* (1860) was published in New York, Torrey correcting the proofs without Chapman's knowing. By the time his *Flora* reached a second edition (1883) Chapman had become an international figure in botany and was receiving letters from European botanists, and this moved him to take up German at the age of eighty. Gray made an excursion with him to see endemic *Torreya* growing on the banks of the Apalachicola.

Another southern "Flora" met with less generosity from Gray. Riddell of New Orleans sent a manuscript in 1851 on Louisiana plants to Henry at the Smithsonian for approval and publication. It was passed on to Gray who dismantled it and filed some of the descriptions in the herbarium at Harvard with specimens Riddell had sent. Riddell eventually published in New Orleans a condensed version with observations by Hale and Carpenter, "Catalogus Florae Ludovicianae" (1852).

A coterie of specialists contributed accounts to the various floristic works of Torrey and Gray, for example: Bebb worked on willows; Thurber, and separately, Vasey, on grasses; and D. C. Eaton on ferns. Torrey and Gray sorted and dispatched incoming specimens for the exploring and surveying expeditions to specialists, edited and organized their contributions, and arranged for the distribution of reprints and gift copies at home and abroad. The correspondence was understandably prodigious. Gray's lack of understanding of the difficulties of wild Texas terrain is shown in his criticism of the ill-fated Berlandier in letters to deCandolle.

Anglo-American rivalries are displayed by a horticultural sensation which flowered at the county seat of Caleb Cope near Philadelphia. W. J. Hooker had sent Cope twelve seeds of the Amazonian water lily, *Victoria regia*, in March, 1851. His gardener John Ellis brought the giant water lily to bloom in August by solicitous care. Robert Buist, the Philadelphia nurseryman, author of one of the several reports of the event, in his *American Flower-Garden Directory* (ed. 6, 1859) was exultant: "the *Victoria* bloomed with more regal grandeur than at any of the Abbeys, Castles, or Palaces of the Eastern world." In 1853, when Meehan, trained at Kew, was Cope's gardener, the 128th flower of *Victoria* opened. Nor were the graphic arts indifferent to the event.

John Fisk Allen published in Boston in 1854 a "brief account of the Victoria regia . . . its discovery and introduction into cultivation. With [six] illustrations by William Sharp from specimens grown at Salem, Mass." This was an elephant folio with sixteen pages of text. It bears very close resemblance to Hooker's handsome folio on the same subject published three years before with four plates in color by Walter Fitch.

Professors Holton of Middlebury College and Orton of Vassar followed the lure of the tropics. Holton, correspondent of Gray, botanized extensively in Colombia, collecting 1800 species, and his narrative *New Granada: twenty months in the Andes* (1857) related "with great spirit," Gray wrote, and "is replete with valuable information." Gray acted as Holton's agent in distributing his *exsiccatae.* Orton, with the blessing of the Smithsonian Institution and the loan of its instruments, made three trips to the eastern Andes. On difficult *jornadas* he collected a few ferns along the Rio Napo and related his experiences in *Andes and Amazon* (1870, etc.) .

Interest in the tropics was aroused by the Darwin and Wallace communications to the Linnean Society. Wallace in turn had acknowledged the influence of a book by an American, William Henry Edwards, *A Voyage up the River Amazon* (New York, 1847) . The collections resulting from such enthusiasm over the tropics stimulated the founding of museums, usually associated with colleges and academies. The Brooks Museum of Natural History, opened in the 1870's on the University of Virginia campus, was typical of many. Emblazoned on its frieze are six naturalists, but no Americans even by adoption like Agassiz or Audubon: Huxley, St. Hilaire, Darwin, Owen, deCandolle, and Lyell. A true representative of the "museum man" was Baird, Assistant Secretary of the Smithsonian Institution, whose zeal carried over to his subscribing funds from his own pocket to support field work by others, and who encouraged the growth of museums in this post-war period.

Increasing student participation in the laboratory was emphasized in the last half of the nineteenth century. Lectures in botany with demonstrations or the filling in of "plant analysis" forms from an examination of local flora specimens, or exhibits of tropisms and simple physiological concepts were increasingly accompanied by laboratory work in anatomy and cell structure. Bessey added laboratory work to his undergraduate botany course at Iowa Agricultural College in 1871, using his one Tolles compound microscope. His *Botany for High Schools and Colleges* (1880) , an adaptation of Sach's *Lehrbuch der Botanik,* reoriented botanical instruction in this country. It transplanted crytogamic botany and physiological anatomy from Europe into American colleges. Graduate studies came more slowly with Harvard's laboratory opening in 1872. By 1885 there were nearly a dozen creditable American laboratories of botany. At Cornell there

was Atkinson, the versatile Leidy at the University of Pennsylvania, and in "the West" at Wabash College in Indiana, Coulter who was to make history later at Chicago. Botanical laboratories opened at University of Michigan, where Gray played a short role, the University of Illinois, Purdue, Michigan State College, Iowa State Agricultural College, and the University of Nebraska. The enactment of the Morrill Act in establishing land-grant colleges in the West was vitally important. Soon there were names in the land. J. C. Arthur, Beal, C. E. Bessey, Burrill, D. T. MacDougal, E. F. Smith, are some who took their doctor's degrees in botany at these institutions before 1900.

The mid-twentieth century with expansion in every facet of living, has brought the Epoch of Government and Foundation sponsorship. Research teams rather than individuals now often distinguish colleges, and they may have not only a coach but a business office, overhead, and dugouts where end runs are planned, with at the end of the fiscal year, a try for a field goal. "Biology is now in one of its happy phases," say Sirks and Zirkle (1964), "but its future is not free from danger. Perhaps the greatest hazard lies in its very opulence. It has already grown far beyond the point where any biologist can master it," and accordingly biologists have divided up the task and have limited individual responsibility. Publications overwhelm. *Biological Abstracts* now publishes some 100,000 abstracts a year, and numbers mount. "There is real danger that if all science continues to grow as it does now, sometime in the near future it may become smothered in its own fat."

The momentum of science is such, says Bentley Glass (1966), that for a very long time the absence of important new theory does not greatly modify the rate of advance. "It is far more a revolution in methods of investigations and in areas of exploration than it is a revolution in our thought. . . . A scientist is scarcely expected any longer to think, only to describe and compute." The carefully planned experiment, like individual data, "must fit into a meaningful pattern of larger investigation which throws light on some major, still-unanswered question of science. The great theory is one that throws a myriad of previously unrelated observations and discoveries into clear relationship."

United States Botany in the Long View:

Botanists were in the beginning naturalists, often actively collecting skins, shells, and insects, as well as seeds. Catesby is remembered for his classic on American birds though he came to collect plants for gentlemen gardeners. Banister, the Bartrams, Rafinesque, and Nuttall, were *naturalists* who made important contributions to botany. American botany was a transplanted science. Foreign-born Nuttall worked for Barton, Parry and Engelmann for Torrey and Gray, Lesquereux **for Sullivant**—all through the nineteenth century American botany

was closely associated with Europe. Then it was the man more often than his works that distinguished an epoch. Though Sloane published a classic on Jamaican natural history, it was his zeal for collecting and fostering collecting among others that gave birth to the British Museum. John Bartram's botanical publications were insignificant, but his stimulus to eighteenth century natural history, by growing, collecting, corresponding, correcting, hosting, prodding, and the role of William (Figure One) in many natural history fields: these lead to the association of his name with the growth of American botany. Barton planned more than he was ever able to complete, but his strategic position in Philadelphia, associated as he was with President Jefferson and thus with government, with the American Philosophical Society, and with scholars of the time, enabled him to *implement* botany.

The impact of German science combined with the rise of the non-sectarian land grant college led to the growth of experiment. What in 1829 were restrictions enforced by the limited nature of our inherent abilities "as if the Creator had said, 'Hitherto shalt thou go but no farther' "—this spirit was challenged in the search for understanding the nature of life. Botany has become in the twentieth century a staff enterprise. From strong personalities, turions of science, who dominated courses of study and gave illustrious names to campuses, departments have emerged with graduate programs, field stations, and patronage. In a very real sense the impersonal staff approach implemented by improved laboratories and microscopes, and esoteric training, have given the rising sciences cumulative momentum. The early graduate student submitted his thesis and took his doctorate abroad, mostly in Germany, though his experiments may have been carried out in this country. He usually returned to direct a department. Typical of the Graduate Laboratory Epoch was B. L. Robinson, an undergraduate at Harvard under the guidance of Goodale and Farlow, he took his Ph.D. degree at Strasburg in 1889, then returned to direct graduate work in taxonomy at Harvard from 1892 until 1925. Another was Heald who took his doctorate at Leipzig in 1897, then directed students in plant pathology at Nebraska, Texas, and Washington State University; Peirce, Ph.D. Leipzig, 1894, went to Stanford in plant physiology. With the maturing of the American university there was encouragement to take a post-doctorate abroad before taking a position in this country: a kind of transition to cope with any lingering sense of inadequacy of an American degree. Now with Commonwealth Fellowships from England, and other like opportunities around the world, promising graduate students come to American universities.

MORPHOLOGY AND ANATOMY

Sherwin Carlquist

(*Rancho Santa Ana Botanic Garden*)

There is no "American School" of plant morphology. The nature of American science makes the development of a concerted or unified program unlikely; individualism tends to prevail. American plant morphology was initiated by, and has frequently been enriched by, immigrants. During its progress, students of plant structure in the United States have been characterized by the wide diversity of their research topics; by their emphasis on rational, broad, and sound outlooks; by their willingness to view plant structure in varied contexts, so that investigations are often interdisciplinary in nature; and by their striving for better techniques improving knowledge. Many American plant morphologists reached this field after working in some other field: morphology does not seem to have been a prime allure in botany. Once entered, however, plant morphology has grasped the imagination and devotion of its practitioners. Readers of Rodgers' *American Botany 1873-1892* may be surprised to find that the chapter entitled "Development of Morphology" is devoted mostly to Asa Gray. Plant morphology as a discipline in the United States, however, began with Jeffrey.

Jeffrey was a native of Canada and his education at the University of Ontario suggests botanical interests not at all for he received his B.A. in Modern Languages and English. His graduate study at Toronto in biology led to doctoral work at Harvard, and after teaching in Toronto he transferred to Harvard where he became Assistant Professor of Vegetable Histology in 1902. This quaint title—later exchanged for Professor of Plant Morphology—indicates recognition, at least in that academic capital, that plant structure had arrived as a legitimate field of interest in the United States.

Darwinian thinking was an early inspiration to Jeffrey's work, and remained an integral element in his career. Intrigued by Van Tieghem's stelar theory, Jeffrey gave this concept an evolutionary dimension, and described the changes in vascular tissue through geological time. A predilection for evolutionary interpretation is basic to Jeffrey's book, *The Anatomy of Woody Plants* (1917). If a desire to reveal phylogenetic alterations in plant structure occasionally led him to speculations which later proved to be unsupported, it also led to a number of ideas so well embedded in our concepts that we no

longer remember them as Jeffrey's. Studies in paleobotany and the formation of coal formed an inalienable part of his work, and those who believe that fossil plants can be understood only when they are studied in structural detail and compared with living plants are able to cite Jeffrey's interests in this regard. Better understanding of living and fossil plants required improvements in microtechnical methods, and Jeffrey contributed a number of techniques. To some of today's botanists, Jeffrey appears to be known by virtue of the solution he formulated for the maceration of woods. Although Jeffrey may not have appreciated it, his main influence may have been in certain students who took training and stimulation from him: I. W. Bailey, Eames, Sinnott, and R. H. Wetmore, to mention a few. Two of these were later to take positions at Harvard, both continuing and modifying the Jeffrey tradition.

The late I. W. Bailey provides United States anatomy with a key figure and a personality as American as Emerson, Twain, or Charles Ives. His boyhood years were spent in the high Andes where his father, a Harvard astronomy professor, was engaged in constructing an observatory at 19,000 feet on the peak El Misti in Peru. This lonely and extreme locality provided Bailey with a stimulation never quite forgotten, for the strange cushion and rosette plants became reminders of some of the intricate pathways and exceptional products of evolution. As Assistant Professor of Forestry at Harvard he paid considerable attention to typical problems of forestry, but his published work at this time reveals his early interest in the basic principles of plant construction. His work on the finer structure of cell walls and pits went to the limits of the capability of the light microscope, and he used ingenious techniques such as artificial expansion of cell walls, crystallization of materials within cell walls, and forcing liquids into woods.

The Bussey Institute, a semi-autonomous Harvard affiliate which granted its own degrees and which employed Bailey, was devoted to research in the applied sciences. Bailey's 142 papers present only a fraction of the original knowledge he developed there on plant anatomy. He served as an aeronautical engineer during World War I, a task in which his understanding of timbers was brought to bear on the problems of airplane construction. Bailey was aware that knowledge of plants in the field enhances understanding of plant structure, provides inspiration and adds a dimension to laboratory work. His travels included Cuba, British Guiana, California, the Carnegie Institution of Washington Desert Laboratory in Arizona, and other parts of the United States. Plant anatomy, however, is mostly a laboratory science, and with equipment which would be considered hopelessly antique by today's standards, Bailey produced preparations and photographs of remarkable clarity and precision.

For Bailey phylogeny was a *leitmotiv* which neither could nor should be avoided. Whether studying the plane of cambial divisions in angiosperms, or the intricate adaptations of ant-plants from Africa, his studies were always conceived as constructs in the service of evolution. His phylogenetic bent, inherited in part from Jeffrey, led him and his students, such as F. H. Frost, Kribs, Barghoorn, and Cheadle, to unravel the fascinating patterns of evolution in treacheary tissue of vascular plants. These studies, so basic to our understanding of plant phylogeny, form a contribution both brilliant and monumental. Bailey comprehended the diversity of wood structure exposed by workers such as Solereder and found in a welter of detail the keys to xylem evolution. He showed remarkable discernment in his selection of length of tracheary elements as a key to evolution of xylem. By using this criterion and finding those features of xylem correlated statistically with it, he and his students created a logically sound framework for understanding wood evolution. Included in this effort were the now-famous papers by Frost on vessel-element evolution, Kribs on ray and axial parenchyma, and studies by Barghoorn which linked ontogenetic and phylogenetic aspects of ray development. In monocotyledons, Cheadle attained a comprehension of and satisfying scheme for understanding of xylem evolution. Although in dicotyledons, vessel elements were introduced into, and specialized in, xylem of roots and stems simultaneously, Cheadle showed an organographic progression in monocotyledons. Vessel elements were introduced into, and specialized within, roots, stems, rhizomes, inflorescence axes, and leaves, in that order. The usefulness of Bailey's concepts is not limited to angiosperms, as recent work on fern tracheids by Richard A. White shows.

Bailey sensed early in his career that "Amentiferae" represent derivatives of various phylads, specialized for anemophily, and that vesselless angiosperms such as *Trochodendron* carried in their internal structure vestiges of the early history of flowering plants. Bailey and his students and co-workers such as Swamy, Charlotte Nast, and Lillian Money contributed many distinguished studies on "woody Ranales." These papers derived from Bailey's conviction that in comparative studies of these genera we may, with care and discernment, find the primitive characters of angiosperms. This significance is also evident in papers by Canright on Magnoliaceae and other genera of dicotyledons rich in primitive characteristics. Others inspired by Bailey who have contributed to this renascence of interest in anatomy of "Ranales" include Thomas K. Wilson (Myristicaceae), Shirley Tucker (ontogeny of flowers in Winteraceae) and Moseley (Nymphaeaceae).

The continued usefulness of studies in wood anatomy is demonstrated by the work of Heimsch, Tippo, and William L. Stern. Problems in identifying materials of Icacinaceae led Richard A. Howard

to a thoroughgoing anatomical study of the family under Bailey's direction. Though he has turned his attention to problems of phytogeography, one of Howard's main interests is in nodal anatomy, and his work has expanded our understanding of this field, one in which Bailey had demonstrated phylogenetic significance.

A comparative anatomist who brought his interests from England to the United States is Tomlinson who had studied with Metcalfe at the Jodrell Laboratories, Kew. Metcalfe had decided to undertake the enormous project of summarizing data on anatomy of monocotyledons, and Tomlinson, working at the Fairchild Tropical Garden in Miami, Florida, has energetically contributed to this work. The volume on palms in the *Anatomy of Monocotyledons* (1961) is his work. He has also produced a cinematic representation of palm-stem anatomy, in which serial sections of palm stems are photographed and viewed in a continuous series, so that a sort of dynamic portrait of vascularization is achieved.

Comparative morphology has offered, it seems, a great temptation to theorizers, but American morphologists have presented very few theories which one may term imaginative inventions at best, outlandish myths at worst. The explanation may lie in the seemingly democratic atmosphere of American science where respect is not maintained by position or seniority, but by the intrinsic value of contributions. Bailey may be taken as the ideal American plant anatomist. With Bailey pressure from colleagues was irrelevant, for with him the scientific method was a matter of conscience and integrity. He felt that his own papers spoke for themselves, and he disliked presentation in book form of material the accuracy of which he could not verify from his own research. He was, however, secretly quite pleased that an assortment of his papers could be presented without alteration in book form as *Contribution to Plant Anatomy* (1954). Bailey is to be admired for his breadth of mastery in various fields: taxonomy, histochemistry, and cytology to mention a few. This breadth was acquired, not from required college courses but through his natural inclination to learn everything he could pertinent to his studies in plant structure and phylogeny.

Eames, contemporary of Bailey and student of Jeffrey, was sparked by plant diversity seen in the field. His experiences in Australia during 1910-1911 were followed by equally stimulating acquaintanceship with floras of New Zealand, Samoa, South Africa, Cuba, Hawaii, and Europe. His enthusiasm for field botany was communicated to students in an effective way, notable for clarity and soundness rather than for rhetoric. Eames was a taxonomist occasionally as in his co-authorship with Wiegand of the fine *Flora of the Cayuga Lake Basin, New York* (1926). Eames and MacDaniels *An Introduction to Plant Anatomy* (1925) filled a vacuum, and gave great stimulus to studies in plant

anatomy. Generations of studies have also attested to the clarity, detail and usefulness of Eames's *Morphology of Vascular Plants, Lower Groups* (1936), which provided the first comprehensive tool for teaching morphology of vascular cryptogams in English. It is appropriate that Eames's interests should have been followed by a student of T. T. Earle at Tulane University, David Bierhorst, whose many excellent studies at Cornell show a continuing American interest in this field.

Although he guided many graduate students, Eames found time for the production of many papers particularly on floral anatomy. He demonstrated an appendicular nature for the inferior ovary in Rosaceae—Pomoideae, as well as in Ericaceae. While over-interpretation of vascularization of flowers can lead to dubious phylogenetic ideas, there are undoubtedly many floral features useful in taxonomic comparisons. Eames was alert to this and has demonstrated many instances of this potential application. Floral vascularization was the field chosen by many of Eames's graduate students, who with other recent workers such as Carl L. Wilson (Dilleniaceae), Barbara Palser (Ericales), and Sterling (Rosaceae), have continued in this field. Eames's varied observations and insights received expression in his *Morphology of Angiosperms* (1961).

Chamberlain obtained his doctorate in 1897 from the University of Chicago and studied in Bonn with Strasburger whose work on life histories offered many precedents. Chamberlain summarized many years of work in his *Gymnosperms, Structure and Evolution* (1934) which has recently been reprinted. His *The Living Cycads* (1919) has similarly been revived for a younger generation of botanists. The career of Buchholz who obtained his doctorate in 1917 from the University of Chicago ultimately led to a professorship at University of Illinois where he centered on gymnosperm embryology. Buchholz was one of many American botanists whose studies led him to New Caledonia and other parts of the Pacific.

The patterns of American pioneering and westward migration led plant morphologists and anatomists to western United States as positions became available, and though they were isolated, their isolation had positive aspects. In newly-opened positions and without the influence of old institutions to shape their careers, they developed distinctive new programs. They were not isolated for long because they soon attracted students and contributed to the growth of major institutions of learning in the West.

One such individual was Adriance Foster who attended Cornell and Harvard where his enthusiasms were aroused by Bailey. After obtaining his ScD. in 1926, Foster accepted a position at the University of Oklahoma, but shortly thereafter became the first plant anatomist at University of California, Berkeley. His early studies on histogenesis of

leaves and bud scales of *Carya* foreshadowed his preoccupation with developmental anatomy, and whether he was analyzing shoot apices of cycads or elucidating the nature of sclereids in *Camellia* leaves, he was exploring ontogeny. His desire to study and collect materials in the field took him to Cuba to study *Microcycas,* to Brazil to study Quiinaceae, and to New Caledonia to study *Boronella* (Rutaceae) and other plants. The clarity and precision of Foster's papers are characteristics also present in his teaching. His *Practical Plant Anatomy* (1942) demonstrates the high value he placed on effective teaching. The result is evident in the many students he guided and the diversity of their accomplishments. Gifford expanded our knowledge of apical organization and co-authored with Foster the widely-used text *Comparative Morphology of Vascular Plants* (1929). Others, such as Warren Wagner, with a particular interest in cytogenetics of ferns, have interested themselves in anatomy of lower groups of vascular plants. Comparative anatomy of angiosperm groups has been the concern of a few Foster students such as Morley, at University of Minnesota, who has studied Melastomataceae. Most of Foster's students have favored developmental anatomy. Boke, for example, has delineated the ontogeny of cactus areoles in excellently-illustrated studies. Other Foster students began by studying problems of apical growth, but soon went beyond a strictly descriptive approach. For them, the transition from developmental anatomy to studies on morphogenesis was a natural and easy one. Ernest A. Ball, for example, influenced by Wardlaw and other students of morphogenesis, utilized experimental techniques in surgically-altered apical regions and isolated cells to demonstrate capabilities of organization.

The early leaders in morphogenesis in the United States were, however, plant anatomists at such institutions as Harvard and Yale. Wetmore, a Canadian schoolteacher who obtained his doctorate at Harvard in 1924, shifted from descriptive developmental anatomy to tissue culture, chemical studies, and other techniques. His infectious enthusiasm helped to popularize morphogenetic studies though his continual self-reeducation limited his published works. Late in his career Sinnott, who began in plant anatomy and early collaborated with Bailey before becoming involved with administration at Yale, published *Plant Morphogenesis* (1960). American studies on morphogenesis have been influenced by such Englishmen and their works as D'Arcy W. Thompson's *On Growth and Form,* and by Wardlaw. Matzke, whose work at Columbia was of morphogenetic nature, has studied problems of cell shape by various devices, ranging from statistical (counting faces of cells in particular tissues) to experimental (duplicating cell populations with such devices as soap bubbles). Morphogenesis grew largely out of plant anatomy and seems to have displaced plant anatomy at many American institutions. Refinements

in study of morphogenetic processes seem to lead ever further from plant anatomy, and to form ever closer links with biochemistry and physics.

Plant anatomy has developed in other directions, however. Katherine Esau and her parents left Russia for Germany during her college years. After graduation from the Agricultural College of Berlin, she began work with the Sloan Seed Company in Oxnard, California, then was employed by Spreckels Sugar Company, in Salinas Valley, where she devoted herself to the problem of breeding strains of sugar beets resistant to the virus causing curly-top disease. Transmission of the virus became an area of potential interest for her and she entered the College of Agriculture at Davis (now University of California, Davis), for a doctorate and remained to become a professor. With plant physiologist Crafts she focused on phloem as the tissue in which virus transmission in plants occurs. Surprisingly little information on phloem was available and her investigations were significant. Working both with other botanists and independently Esau demonstrated an exceptional ability for attacking basic problems. Her papers are notable for textual clarity and illustrational excellence. She perfected a specific stain for phloem and with her co-workers devised delicate experiments with preparations of living phloem to determine whether sieve-tube elements, although enucleate, have differential permeability. Her studies on histology were always accompanied by ontogenetic studies. Thus she showed that sieve-tube elements in the grapevine can be reactivated for a second year of function. Almost all her studies naturally involve crop-plants, for example, her studies on collenchyma ontogeny were done on celery, and her studies on vascular differentiation utilized flax. Her enduring classic, *Plant Anatomy* (1953) now in its second edition, and *Anatomy of Seed Plants* (1960) broadened her entry into English-speaking classrooms. The developmental aspect of her studies matured into *Vascular Differentiation in Plants* (1965). Her early interest in virus transmission continued with *Plants, Viruses, and Insects* (1961).

Because Davis campus did not tend to attract students in botany as did Berkeley, Esau did not have many graduate students. As those who have heard her speak know, she possesses an elegant lucidity in oral expression which matches her written style. Her election to the National Academy of Sciences in 1957 was a well-justified form of recognition to a distinguished career. As retirement approached she turned to the complexities of the electron microscope. Her ultrastructural work is particularly successful. She has analyzed both developmental and mature states of conducting tissues; she has studied cells not in isolation, but in sections where the orientation of cells is intact; and she has bridged the gap between the electron and light microscope so that the results of one are congruent with the results of the other.

Flora Murray Scott also adopted California. Born in Scotland, she received her doctorate at Stanford University and was the first plant anatomist at University of California, Los Angeles. She has enjoyed exploring forgotten or overlooked corners of plant anatomy and cytology. Her zest has led to publications on such curious phenomena as water-storage tissues of ocotillo (*Fouquieria*), the pulsating movements of the living nucleus, and the nature of lipids in the cytoplasm. On her retirement she also anticipated exploration with the electron microscope.

Johansen also did graduate work at Stanford University, but speech and hearing difficulties prevented his entering an academic position. His *Plant Microtechnique* (1940) shows the breadth of his experience in the field. Many of his preparations have involved embryology in gymnosperms and angiosperms and led to his *Plant Embryology* (1950). Although his teacher, Douglas Houghton Campbell, was interested in embryos of vascular cryptogams, Johansen's work represents more his own initiative than the continuation of a tradition.

One California worker, largely self-trained as were most American plant anatomists, Marion Stillwell Cave entered graduate work in genetics because a scholarship happened to be available. Aiding Babcock during his famous *Crepis* research, she became skilled and soon excelled in cytology. Her papers on embryo sacs of Liliaceae and the taxonomic use of embryo-sac characters have received recognition.

Another "de novo" field of morphology was developed by Wodehouse, born in Canada and with an A.M. degree from Harvard, who studied the Compositae. His "The Phylogenetic value of pollen-grain characters" (1928) demonstrated that in *Vernonia* distinctions in exine patterning can be criteria for recognition of subgenera and species. The grouping of spines into ridges and allied sculptural features prove good indicators of phylogenetic specialization. Mutiseae were studied with equal success. Although his employment was non academic, Wodehouse's *Pollen Grains* (1935) was the first full-length treatment of pollen morphology and its relationship to taxonomy and other fields. His work undoubtedly did much to create the popularity which palynology now enjoys although the more recent studies of Erdtman are more detailed. The exquisite illustrations of surface sculpturing in Wodehouse's work have never been equalled.

Pollen morphology in relation to taxonomy has been further pursued by Dahl of the Morris Arboretum, and Walter H. Lewis of the Missouri Botanical Garden. The electron microscope naturally plays an important part in the shift toward greater detail: Rowley, University of Massachusetts, is a recent worker on families of monocotyledons. The success of the electron microscope in a comparative way on pollen has been demonstrated by Skvarla whose studies on Compositae pollen are an ultrastructural extension of Wodehouse's work.

Electron microscopy of plant tissues received its greatest impetus, probably, from the work of Europeans such as Frey-Wyssling. The pace of electron microscopy accelerated as American laboratories became equipped and contributions are too numerous, directions too varied, and works too many to chronicle here. Almost every aspect of ultrastructure is currently under investigation by productive workers.

The earlier work in electron microscopy featured establishment of basic structural principles of cell walls, chloroplasts, mitochondria, etc. With clarification of these components, ultrastructural studies seem destined to explore developmental aspects and macromolecular elements. Comparative studies will be pursued as the use of the electron microscope widens.

As American morphology shifts toward studies in morphogenesis and ultrastructure, certain traditional fields in plant anatomy may dwindle, but such fields will doubtless revive from time to time as enterprising Americans retrain and equip themselves whether or not there is public recognition of their work. And as in the past, American morphology and anatomy will benefit from overseas papers, or colleagues who come to visit or to stay, and who demonstrate that studies on plant structure are an international concern.

PLANT GENETICS AND CYTOLOGY

Conway Zirkle

(*University of Pennsylvania*)

When English settlers landed in Virginia and Massachusetts, they were faced with an unfamiliar climate and with soils unlike any they had known. Crops they planted from seeds of the Old World did not thrive in the new environment. The settlers were able to survive only through food they could buy or steal from the Indians. Friendly Indians also taught them how to grow American plants and, once this was learned, the settlers were no longer threatened with starvation. They adopted Indian agricultural methods, but experimented widely—sometimes desperately, since they were forced to rely on the native American plants. In adopting American plants and adapting European plants to American conditions, they unknowingly subjected their crops to both artificial and natural selection and at times hybridized their plants without knowing exactly what they were doing.

The American Indians were skilled agriculturists. Indian agriculture was based on the hoe, European and Asiatic on the plow; American cultivated plants could be and were cultivated effectively as individual plants while Old World crops could best be grown in plowed ground. Even as late as the eighteenth century settlers used the hoe technique when they raised Indian plants and plows were not widely used on farms in Virginia and Maryland. Such American plants as Indian corn, tobacco, potatoes, and squash could be grown effectively in hills and weeds eliminated by hoeing. It was not until the nineteenth century that Americans were well equipped with plows.

Even today about one fourth of the crop acreage in the United States is devoted to American plants, and many of these cross spontaneously with their relatives and the hybrids they form are not hard to recognize. It is not an accident that the first plant hybrids known to be hybrids were of American plants and were first recognized in America. Before they were identified, the very practical settlers had selected some fortunate hybrids for planting and thus were increasing their agricultural yields.

The first description of a hybrid plant as a hybrid plant is somewhat ambiguous, and not of a cultivated plant. It is brief and to the point. English-born John Lawson in *A New Voyage to Carolina* (1709) wrote: "Bastard Spanish is an oak betwixt the Spanish and the red oak; the chief use is for fencing and clap boards. It bears good acorns."

Bastard Spanish is a naturally occurring hybrid between *Quercus falcata* and *Quercus rubra*. Did Lawson know that Bastard Spanish was a hybrid? In German whenever a bastard was not a human bastard it was a hybrid—a bastard plant was a hybrid plant. This passage was translated into German and published in Bern, Switzerland, in *Neugefundenes Eden* (1737). Here it reads: "Bastard Eiche ist ein Eiche Zwischen der Spanischen und Rothen Eiche / das Holz is gut für Fasstauwen." Here a hybrid oak is labeled hybrid. Now when the English called a plant a bastard they merely meant that they considered it inferior. Later, when they learned of hybrid plants they called them "mule plants." Hybrid plants were knowingly called "bastards" in the United States in 1755, 1801, etc.

The first unambiguous account of plant hybridization is also American. In a letter dated 1716 the New England divine, Cotton Mather, wrote (Zirkle, 1935):

"In a Field not far from the City of *Boston,* there were lately made these Two Experiments.

"First: my Friend planted a Row of *Indian Corn* that was Colored Red and Blue; the rest of the Field being planted with corn of the yellow, which is the most usual colour. To the Windward side, this Red and Blue Row, so infected Three or Four whole Rows, as to communicate the same Colour unto them; and part of ye Fifth, and some of ye Sixth. But to the Lee-ward Side, no less that Seven or Eight Rows, had ye same Colour communicated unto them; and some small Impressions were made on those that were yet further off.

"Secondly: The same Friend had his garden ever now and then Robbed of the Squashes, which were growing there. To inflict a pretty little punishment on the Thieves, he planted some Gourds among the Squashes, (which are in aspect very like'em) at certain places which he distinguished with a private mark, that he might not be himself imposed upon. By this method, the Thieves were deceived, & discovered, & ridiculed. But yet the honest man saved himself no squashes by ye Trick; for they were so infected and Embittered by the Gourds, that there was no eating of them.

"Several Useful Hints relating to Vegetation and Agriculture, may be taken from these Experiments; which I wholly leave to your Sagacity, and that of ye Philosophers to whom you may see cause to mention them."

Different colored grains on an ear of *Zea Mays* had been recorded by Tabernaemontanus as early as 1588 and described many times by others, but the cause of the mixtures was not stated until Mather ascribed it to cross-pollination. The very next year, 1717, a hybrid *Dianthus* was described in Britain, a hybrid cabbage in 1721, and a hybrid *Mercurialis* in France in 1719. In 1724 a second record of the hybridization of color varieties in *Zea* came from Martha's Vineyard,

an island off Massachusetts. Paul Dudley noted the mixing of colors even though the parental plants were separated by a ditch of water. Which meant that the mixture could not come from the rootlets fusing underground, but had to come from cross-pollination. In 1735, James Logan, Governor of Pennsylvania, published his experiments on pollinating *Zea,* and this work helped to establish the fact that plants reproduced sexually.

Many experiments in hybridizing plants were now being made in Europe and continued in America. John Mitchell in Virginia wrote on plant and animal hybrids (1738) ; and John Bartram of Pennsylvania described a hybrid lychnis (1750). Charles Alston, Professor of Botany at Edinburgh, however, attacked the idea of sex in plants. He was answered by "two celebrated botanists of North America" in letters to *The Gentleman's Magazine* (1755). "J. C." (John Clayton?) told of "bastard or mule plants" and described a hybrid cypress vine, *Ipomoea quamoclit.* "J. B." (John Bartram?) told how cucumbers, squashes, melons and gourds would hybridize spontaneously if grown near each other.

In 1760 Linnaeus published his famous *Disquisitio de . . . Sexum plantarum* and 1761 Koelreuter his *Vorläufige Nachricht von einigen das Geschlecht der Pflanzen.* From then on experiments in plant hybridization became a professional occupation. Plant hybrids would no longer be recorded as mere natural curiosities. During this period corn in America was being hybridized continually but not always intentionally. Although small grains were now being grown successfully in America, corn was still the chief crop. Cultivated corn was by this time a hybrid mixture of northern flint with southern "gourd seed corn." As early as 1705 Robert Beverly of Virginia had recorded a grain of flint corn on an ear of "she" corn, as "dent" was called at the time, and here we have observational evidence of how early crosses were made, but most hybridization went unrecorded.

Joseph Cooper of New Jersey, in a letter dated 17 April 1799, told how he had greatly increased his yield of Indian corn by selecting superior strains from variable hybrid progeny. By carefully selecting desirable qualities, he had improved his potatoes, radishes, squash and asparagus. He was also able to acclimate southern watermelon to the New Jersey climate by collecting seed from those that ripened first. During the whole of the eighteenth century Americans had been developing many new strains—many new varieties of apples, pears, peaches, plums, and cherries. Derived chiefly from imported European varieties they were soon reputed to be superior to their Old World ancestors, and the high quality of American grown fruit was commented on by many immigrants. European imports, of course, were crossbred, mostly accidentally. As a rule farmers raised many different types for home consumption and, at first, propagation was by

seedlings. Inferior trees naturally were cut down and, later on, superior trees were reproduced vegetatively. As the frontier moved west the settlers carried with them seed from favorite varieties and the process repeated. The end results were excellent. The procedure was purely empirical since there was little or no interest shown in the theory or principles of plant hybridization. Toward the end of the nineteenth century however there developed an intense interest in hybridizing fruit trees, both the imported and native American, and thousands of crosses were made. The motive was still primarily practical—to develop superior varieties.

The fruit which had promised most to the settlers actually gave the least. American grapes were a disappointment and *Vitis* presented the settlers with their most complicated horticultural problem. Some twenty species of grapes are native to the United States and the country is obviously good grape country. It was hoped that wine made from American grapes could be exported and become a major source of revenue, but most native grapes are too harsh for wine making although many are excellent table grapes. Without exception they have too much flavor and too little sugar for wine. Obviously the long-cultivated and expertly-selected European wine grape, *Vitis vinifera,* should be planted and wine made from it, therefore many varieties were imported, but always "a sickness fell upon the vines" and they died. European grapes were susceptible to phylloxeran root parasite to which American grapes were immune. The few Americans who really knew how to drink wine had to import what they drank since they were unhappy with wine made from the Scuppernong (*Vitis rotundifolia*) and from the Catawba. By the early nineteenth century, however, American grapes had been widely hybridized *inter se* and most of them had also been crossed with European: more than forty of the cultivated American grapes are hybrids. These are the raw material for modern New York State wines which, by modern methods, are very pleasant and stimulating. California wines are made from European grapes.

The spread of small grains toward the western frontier and ultimately to the wheat belt, gave rise to a number of problems that were solved empirically and generally unknowingly by the migrating farmers. Wheat cultivation for example was extended to a region of relatively little rainfall and into northern states where winters were too severe for winter wheat and the growing season was short. The wheat seed was a mixture of many kinds and the selection that occured in the marginal region was drastic and effective. The farmers usually assumed that wheat, if given time, could adapt to harsher climates, and they paid little attention to the method of adaptation. In 1858, the year in which Darwin and Wallace described natural selection, Klippart of Ohio showed (independently of Darwin and Wallace)

how natural selection operated within a cultivated species: "The more prolific varieties of wheat when mixed with the less prolific, bringing forth proportionately more grain, will in a few successive crops give a preponderance of the more prolific varieties and in the end supersede entirely the others and thus, without one variety being transformed into another, the character of the crop may be entirely changed in a few years."

During the second half of the nineteenth century numerous Agricultural Experiment Stations were established and the genetics of cultivated plants was studied intensively. It was a premendelian genetics yet a number of advances were made. American universities at this time contributed little to the advancement of plant genetics yet an isolated instance gives our first real recognition of the problems involved in xenia. Xenia is the name given to the phenomenon in which not only the embryo but also the endosperm, which seems purely of maternal origin, shows the effect of foreign pollen. The direct effect of foreign pollen on grains of corn had been noted in 1716, but this seemed only natural until it was realized that plant embryo and endosperm were two separate and distinct tissues.

When Focke coined the term xenia in 1881 he included in the term much that was later called metaxenia. The direct effect of pollen upon the endosperm was not explained until 1899, when Navaschin described the double fertilization, whereupon both de Vries and Correns hastened to publish their papers on hybridization in *Zea mays*. From the 1860's on, several European botanists, e.g., Vilmorin (1867), Hildebrand (1868), Koernicke (1876), etc., had noted the effect of foreign pollen on the endosperm of *Zea mays* but, of course, could not explain it. In 1858 Asa Gray had explained the occurrence of different colored grains on a single ear of corn in two ways, (1) as due to cross-pollination in the previous year and, (2) as due to the direct effect of pollen on the endosperm. This may well be the first recognition of the real problem of xenia.

Research in plant genetics carried out in the various state agricultural experiment stations had a strictly practical orientation. The basic problem was to increase the yield per acre in the ever expanding farm lands. The importation of foreign strains and the development of superior domestic strains were the obvious means of increasing agricultural production, as was the search for races adapted to the many different agricultural regions. Of practical importance also was the discovery of races immune to always threatening fungal diseases. The development of new races proceeded rapidly and, in wheat alone, over 350 pure bred lines had been created by the twentieth century.

Pure bred varieties were preferred by the farmers because their qualities and breeding properties were stable. Their yield per acre also turned out to be much above that of the earlier grown unpedigreed

stocks. On the other hand, the increased yield of certain hybrids was recorded many times but was not understood. This was an intriguing problem and an intense investigation into hybrid vigor was begun. No explanation for it was discovered until well into the twentieth century.

Hybrid vigor had, in fact, been utilized ever since the beginnings of agriculture. Even in classical times, the better varieties of fruits were accidental hybrids which would not breed true from seed, but which could be propagated by cuttings and grafting. At the beginning of the nineteenth century, Knight had noted the hybrid vigor that followed the crossing of varieties and this vigor was recorded many times in the botanical literature of Europe. Mendel himself noticed it in his classic paper, and Charles Darwin devoted a whole book, *Cross and Self Fertilization in the Vegetable Kingdom* (1876) to the subject. Darwin's book stimulated American experimentors and Beal (1880) described how he had increased the yield of corn by 50% by mixing two stocks of the same variety. He was followed and his work confirmed towards the end of the century by W. M. Hays (1889), J. W. Sanborn (1890), S. W. Johnson (1891) and McCluer (1892). H. J. Webber (1900) found his hybrid corn plants to be 50% taller than their open pollinated parents. No valid explanation of hybrid vigor was offered, however, before Mendel's paper was recovered, and no way had been found to stabilize hybrid yields.

This concentration of research in plant genetics in the agricultural field is exemplified by the last pre-Mendelian conference. Of the nine Americans who attended the 1899 Hybrid Conference of the Royal Horticultural Society only four had university connections, and only one of these was in a department of botany. The other three were from horticultural and agricultural departments.

The discovery of Mendel's overlooked paper in 1900, completely reoriented research and plant and animal genetics were unified—into *genetics* — and completely internationalized. American universities quickly adopted genetics as a new and important discipline and included courses in their botanical and zoological departments or established new departments. Theoretical advances now came from the universities and progress in the subject as a whole was exceptionally rapid. The outbreak of World War I, however, made genetics for a time almost an American science.

The first great post-Mendelian advance, or "breakthrough" came from Morgan and his group at Columbia University. Specific genes were finally located in identifiable chromosomes, linkage between genes put on a quantitative basis, and the picture of the machinery of heredity began to appear in detail. The chromosomes of *Drosophila* were mapped, genes assigned in their linear order, and later, when the salivary gland chromosomes were investigated, genes were located in

visible structure within the chromosomes. Chromosome mapping was soon extended to a number of plant genera, most completely to corn chiefly through the work of Emerson and his students at Cornell.

Native American plants turned out to be especially favorable for genetic research. Chromosomes of tomato have now been extensively mapped. Professor East of Harvard and his students made extensive crosses in *Nicotiana* and were able to identify certain genomes that persist through several sister species. In Indian corn the greater number of both theoretical and useful discoveries were made. East and Emerson crossed a short eared pop-corn (mean length of ear 6.6 cm., coefficient of variation 12.3%) with a long eared sugar-corn (mean 16.8 cm., C.V. 11.1%). The ears in the first hybrid generation had a mean length of 12.1 cm., and a coefficient of variation of 12.5%. In the second hybrid generation the mean length of the ears (12.6 cm.) was almost the same as the first but the coefficient of variation had increased to 22.3%. The increased variability in the F_2 generation was due to a Mendelian segregation and in this generation both parental types reappeared. Thus size or quantitative heredity was shown to be Mendelian.

Economically, the most important practical advance made by American plant genetics was the creation of hybrid corn. The advent of Mendelism gave a clue that made it possible to stabilize hybrid vigor and retain the advantages of what came to be called heterosis. Stabilizing the vigor and genetic interpretation of the factors involved was worked out independently by East and Shull (1908) and carried on by D. F. Jones (1917).

During the period between Mendel's publication and its discovery, great progress was made in knowledge of the cell and of cell components, but this knowledge was not applied to problems of heredity although there was speculation as to which parts of the cell were responsible for the transmission of hereditary characteristics. Several investigators suspected that chromosomes were the bearers of heredity and in 1892 Weismann stated definitely that they were. In 1903 Sutton of Ontario, Canada, reported that chromosomes passed from generation to generation precisely as did Mendelian factors, and that chromosomes carried the factors, later called genes. Thus genetics and cytology came together and the active field of cyto-genetics was born.

Americans made almost no contributions to nineteenth century cytology which was predominately a German science, but Americans who had studied in Germany brought back a knowledge of the subject and of research methods developed in Germany. It was after Mendel had been discovered that Americans made contributions. Sutton's work has been referred to. McClung (1902) of the University of Pennsylvania discovered the X-chromosome in the males of Orthoptera and E. B. Wilson (1905) of Columbia found that females had two such

chromosomes and the male but one. Thus was the basis for determination of sex discovered. In 1910 Morgan located specific genes in the X-chromosomes of *Drosophila* and Emerson and his students diagrammed them in the chromosomes of Indian corn. Genetics was advancing rapidly: the physical basis of genetics had been found. The genes had been located.

In the twentieth century cytology developed rapidly through the cumulative effects of innumerable minor contributions made both by botanists and zoologists in many different countries. From research that centered in America we may begin with polyploidy. One American genus especially suited for this type of investigation was *Nicotiana*. This work, summarized by East (1928), shows how the study of chromosomes makes it possible to discover the relationship of the several species, and thus to trace the evolution within the genus. Here the chromosome numbers in the species range from eight to 48. Polyploidy in *Crepis* is even more spectacular. Babcock (1947) found that in 78 species he had investigated, the chromosome number ranges from six to 88. The chromosome sets were highly individualistic and specific sets of chromosomes (the genomes) could be identified in a number of different species. Here the role of plant hybridization in evolution could be evaluated. By studying the chromosomes of *Ranunculus* and its relatives, and in identifying the genomes, Gregory (1941) was able to reorganize the systematics of the family.

The effects of duplicating individual chromosomes in *Datura* were described by Blakeslee and his co-workers. *Datura stramonium* normally has just twelve pairs of chromosomes. Each of the twelve chromosomes has been duplicated so that it was possible to obtain twelve different trisomics. Thus the genetic effects of an extra chromosome from each of the pairs could be recorded. Many other such instances could be cited.

The puzzling behavior of *Oenothera lamarckiana,* which frequently produced plants which were identified as belonging to other species, led de Vries to his theory of mutation. For some forty years this and other species of *Oenothera* were investigated intensively by several botanists. *Oenothera lamarckiana* behaved as if it were an almost permanent hybrid. Finally Cleland (1924, et. seq.) solved the problem of its peculiar behavior. In this genus crossing over had occurred between non-homologous chromosomes so that in the reductional division the chromosomes formed rings and alternate chromosomes went to each of the daughter nuclei. Here the assortment of chromosomes was not random, and a pair of balanced lethals in a ring would eliminate the homozygous segregants. The heterozygous segregant or hybrid would thus breed true and only the very rare rearrangements would also survive. It was natural that these new forms would have been identified as mutants.

The role of chromosomes in determining sex in animals was discovered very soon after Mendel's work was recovered. The role of the chromosomes in dioecious plants is more complex, but when polyploidy can be introduced into such plants it becomes possible to evaluate the role of chromosomes in sex determination a little more precisely. It is here that the role of the Y-chromosome was first evaluated.

In 1940 Warmke and Blakeslee, and in 1946 Warmke, determined the inheritance of sex in diploid and tetraploid *Melandrium*. In this genus the Y-chromosome is necessary for the male elements to develop. In diploids, triploids and tetraploids, a single Y-chromosome would produce a dioecious male even in the presence of two X-chromosomes. In a tetraploid that also has four X-chromosomes, however, a single Y-chromosome makes the plant monoecious. Here the role of the Y-chromosome is different from its role in *Drosophila* where sex is determined by the balance between the X-chromosome and the autosomes. Later it was discovered that the Y-chromosome is necessary for the production of maleness in mammals.

The fact that many species of plants cannot be fertilized by their own pollen was recorded by Koelreuter (1764). The genetic solution of self-sterility, however, was not found until 1926 when East and Mangelsdorf solved the problem in *Nicotiana*, and Filzer and Lehmann in Germany solved it in *Veronica*.

Another American contribution to genetics is one that has had both theoretical and practical consequences. Muller in 1927 produced artificial mutations in *Drosophila* by means of X-rays and in 1928 Stadler used X-rays to produce mutations in *Zea mays*. This technique lead indirectly toward a new method of biochemical research. In 1941 Beadle and Tatum published their investigations on *Neurospora crassa*, a fungus whose mutants could not grow on synthetic media unless some amino acid or vitamin had been added. Thus biochemical consequences of the mutations could be identified, and the artificial production of mutants produced a powerful biochemical tool.

The electron microscope has contributed a major "breakthrough" in cytology, just as the Watson-Crick (1953) discovery of the structure of desoxyribose nucleic acid had in genetics. All earlier work is now referred to as coming from classical cytology and classical genetics. Bacteria are a favorite subject of the newer research, and bacteria are still classified as plants. Few botanists have claimed jurisdiction over the viruses although some viruses have been caught carrying genes from one strain of bacteria to another. All of this work is very recent, however, it is not historic—and we might as well stop here, because we have to stop somewhere.

PLANT PHYSIOLOGY

Sterling B. Hendricks

(U.S.D.A., Soil and Water Conservation Research Division, Beltsville, Maryland)

Advanced education and research did not get underway until the 20th century in the United States. Work in plant physiology was delayed even more—in fact, this XI Botanical Congress is about the semi-centennial. Classical work by European physiologists in the last half of the 19th century had no counterpart here.

Joseph Priestley spent only the last ten years of his life in America, coming in 1794 as a dissident minister more interested by then in the freedom of man's mind than in the life-giving qualities of oxygen. He settled away from the center of learning. Philadelphia, and declined an offer to become a professor of chemistry at the University. He did build a pneumatic laboratory which is still preserved with its instruments as a museum in Northumberland, Pennsylvania. One of the American Chemical Society's chief awards is the Priestley Medal.

George Washington and many other of the nation's founders derived their wealth from tobacco, rice, and other exports. The early settlers had brought their knowledge of agriculture from a country where the presence of farm animals had been essential for the constant manuring of the land. Fitzherbert's *The Boke of Husbandrye* (1523), the first printed book of agriculture in England, showed that the value of manuring was fully appreciated by that time. With relatively few farm animals in the early years the settlers eagerly adopted the Indian custom of planting "a fish in every hill of corn." It was easier and less costly to abandon a piece of land, or, as Jefferson was wont to do, plant it to clover for a year or two, and to clear and put a new area into cultivation. Edmund Ruffin, a Virginia planter of the early nineteenth century, disturbed by the results to the soil of constant tobacco planting, and influenced by Humphrey Davy's *Elements of Agricultural Chemistry* (1815) began to experiment with marl of different quantities on his fields. His *An Essay on Calcareous Manures* was published in 1832 with 242 pages and grew to 493 pages in the fifth edition (1852).

Agriculture was aided by granting of federal funds to each state in 1862 to establish colleges of mechanic arts and husbandry. Some of these were complete universities. Physiology started in them after 1900—California, Connecticut, Cornell, Nebraska, and Rutgers—and

in the endowed advanced schools, Chicago and John Hopkins, as well as in the U. S. Department of Agriculture which was also founded in 1862.

The central interest in agriculture and the continental climate of the United States focused early attention on the requirements of plants for water. This took the form of interest in water supply and loss and in root development. Findings of Briggs and Shantz (1911 et seq.) on wilting of plants (the wilting coefficient) and the potential of water in soils were fundamental developments. Briggs, a physicist, later became the Director of the National Bureau of Standards while Shantz, at the University of Arizona, maintained an interest in the water economy of xerophytes. MacDougal and Livingston of the Carnegie Institution's Desert Laboratory, which was established in Tucson, Arizona in 1903 and discontinued in 1941, also worked on evapotranspiration, particularly under desert conditions. MacDougal wrote one of the early textbooks on plant physiology by a United States author (1901). Clements, working at the Desert Laboratory, studied the dependence of root activity on soil aeration (1920) to which attention had been drawn as early as 1904 by the soil scientist, E. Buckingham.

Anyone now interested in root distribution in soil has much to learn from the lifelong work of Weaver which he summarized in two books (1926, 1927) on root development of field and vegetable crops. W. A. Cannon and E. C. Miller, working chiefly with desert plants and those of the prairies, also showed the great ramification of root systems as an important factor in water gathering.

Soil fertility as an evident factor in crop growth led to establishment of several "perpetual" field plots starting in 1882 at the Pennsylvania State University and including plots at Universities of Ohio, Illinois, Minnesota, and Missouri. These were fashioned after the Rothamsted plots to assess the necessity of replenishing common nutrient elements. Ten of the essential elements were known by 1900 and the rudiments of solution culture, based on the work of Knop in Europe, were established. Refinements of the compositions of these solutions, attention to osmotic requirements, and seeking of elements essential for growth were destined to be the central theme of plant physiology for the next 40 years in the United States. During this period, Tottingham (1914), Shive (1915), Trelease (1917), and Hoagland (1920) formulated solutions that have come into wide use. Hoagland came gradually both to a deeper appreciation of nutrient balances in plant growth and to the design of experiments to follow salt uptake upon which an understanding of mechanism might be based. He and his students established the essentiality of several elements and showed the great value in the use of tracers (Rb for K, and Br for Cl) which led to the eventual wide use of isotopic tracers with detached roots in short-time

experiments. Over this same period of 1900 to 1940, knowledge had grown apace both in the United States and Europe on nutrient deficiency symptoms as indices of element essentiality for growth.

Recognition of broad principles and essays on causation were rare in plant physiology in the U. S. as well as the world between 1900 and 1930. Four come to mind. In addition to that of Briggs and Shantz, the list for the U. S. includes the discovery of photoperiodism by Garner and Allard (1920), the crystallization of urease by Sumner (1926), and the observations of Osterhout (1910 et seq.) on membrane behavior. The last of these gave an insight into the manner of salt transport into cells and laid a basis for much of what was to follow on nerve impulse transmission as well as in plant nutrient uptake. The crystallization of urease from Jack bean meal was the forerunner of an eventual epoch in which plant physiology was to become closely allied to biochemistry. It settled the discussion centering on the protein nature of enzymes. The discovery of photoperiodism was more in the spirit of the great discoveries in Europe of the 19th century. It was hardly in accord with the ideas of the times, but rather gave a first sensing of the interaction of plants with their environments.

The details of this initial period of plant physiology in the United States are well summarized in *Plant Physiology* by E. C. Miller (1931), and "Growth of Plants" by Crocker (1948).

THE MIDPERIOD:

A new epoch came in the decade of 1930 to 1940—the epoch of developing knowledge about photosynthesis and hormone action. It started with observations by W. A. Arnold and the late Robert Emerson (1932) at Stanford University on action spectra for photosynthesis and the concept of the photosynthetic unit. It included the comparative aspects of photosynthesis in purple bacteria and higher plants as leading to reductive systems as formulated by van Niel (1931) of the Hopkins Marine Laboratory of Stanford University. In the same period, F. W. Went came to the United States and continued fundamental work on auxin as a plant hormone with associates at the California Institute of Technology. Hoagland's work on plant nutrition was at its peak during this decade. The promise of Sumner's work on urease, however, lay in abeyance as an aspect of plant physiology which rather depended on a more complete development of biochemistry which was in progress. Biochemistry applied to plant physiology was mostly concerned with sugar and starch chemistry and nitrogen metabolism with a view to assessing the states of metabolic reserves in plants. Warburg manometric methods came to be used widely in overall measurements of metabolic rates. ATP, which was isolated by Fiske and Subbarow in 1929, was recognized by F. Lipmann as acting in energy transfer. Identification by Beadle, Tatum and

their coworkers of "one gene—one enzyme" based on work with *Neurospora crassa* gave a secure basis for following metabolic pathways, as illustrated by the course of tryptophane synthesis.

The years of World War II—1938 to 1945 were as intellectually blank in plant physiology as they were morally reprehensible for the world as a whole. A useful byproduct of nuclear research effort during that period was an immense productive capacity for radio-active elements suitable for use as tracers. Availability of C14 and P32 in particular were the basis for new departures in physiology. Calvin and his associates at the University of California brought the full power of C14 use to bear on the chemical pathways of CO_2 fixation and elaboration of sugars shortly after the war. P32 served both to follow the relationships of the phosphorylated sugars that entered so centrally in plant metabolism, and inorganic phosphate in its passage into roots with the early formation of ATP.

The Recent Period:

The years since World War II have brought a developing understanding of numerous facets of plant physiology. There are far more participants in the United States than in early periods. Grouping into teams and the rapidly shifting scene are causes for not mentioning individuals. Rough estimates (in parentheses) are made of the number of research workers active in each subject.

One of the subjects particularly associated with the United States is the course of nitrogen fixation in plants (50). Conditions were found in 1960 for fixation in cell-free solutions prepared from *Clostridium pasteurianum*. This work led to the discovery in 1962 of the non-heme iron-containing protein ferredoxin as the strong reductant involved both in nitrogen and in photosynthesis. Hydrogen was early (1938) found to be inhibitory to the fixation process of nodulated red clover indicative of involvement of an hydrogenase. The recent findings (1966) that the nitrogen reducing system also catalyses acetylene reduction greatly facilitates detection of fixation. Through its use, nitrogen fixation has been found in cell-free extracts of bacteroids from soybean root nodules. Steps in nitrate reduction and the action of nitrate reductase have also been worked out.

Much interest has centered in hormone action implied in aspects of growth (100). This interest received its initial impulse from work on auxin in the 1930 decade. The possible site and nature of auxin action remain active areas of research with current encouraging results. Kinetin was recognized (1955) from requirements of tissues for growth in culture. Its isolation and synthesis encouraged other work leading to recognition of several kinins and to their interaction with auxin in mitosis and growth of callous tissue.

Eventual awareness of the Japanese work on gibberellins resulted in

much work on the variety of physiological displays induced by gibberellic acid and to interest in the manner of action. A recent finding is the enhancement of alpha amalyase activity by GA_3 in barley endosperm tissue of germinating seeds. A fourth class of plant hormones interacting with gibberellins and auxins is represented by abscisic acid (1963) which was recognized as involved in leaf abscission and bud dormancies. Abscisic acid was isolated and synthesized independently by groups in the United States and in England.

Ethylene has long been recognized as being highly active both in plant growth and maturation. Its bearing on after-harvest physiology of fruits stimulates continued work. Interest also is devoted to the manner of ethylene production from methionine in plants.

Phenoxyacetic acid, its derivatives and related compounds were recognized by P. W. Zimmerman at Boyce Thompson Institute in the 1930 decade as having some auxin-like activity when used in low concentrations. In the period between 1940 and 1945, several research teams in the United States and in England recognized that the phenoxy acetic compounds such as 2,4 dichlorophenoxy acetic acid are selectively toxic to plants. The level of toxicity was found to vary greatly among species. This led to immense development of herbicides and to basic changes in the pattern of American agriculture. Physiological interest centered on assays for activity, screening of possible active compounds, and on finding specificities of action on plant species. Phenyl and chlorophenyl carbamates were recognized as useful herbicides in the period of 1945 to 1950. In the next decade, phenyl and chlorophenyl ureas, s-triazines, and quarternary ammonium compounds were added to the growing list of effective substances.

The intense effort to screen possible compounds for herbicidal activity and to reduce the discoveries to practical use has in recent years been accompanied by work on the manner of action of specific compounds. The s-triazine compounds are found to act as enzyme inhibitors. Thus, 3 amino 1,2,4 triazole (amitrol) is inhibitory to one enzyme involved in histidine synthesis in some heterotrophic organisms. The substituted phenyl ureas (diuron, etc.) act as inhibitors of photosynthesis through inhibition of electron transport.

The Weed Society of America which was formed in 1956, now has more than 1200 members. *Weeds* is the official journal of the Society. The chemical industry devoted to herbicide synthesis and formulation continues to grow. More than 100 million kilograms of herbicides are now produced in a year for use on 50 million hectares of land.

Work on tissue culture was started in the United States in 1907 by Harrison of Yale University on neuroblasts of frogs following the initial research in Europe by Haberlandt. Culture of plant organs and tissues was initiated by W. J. Robbins in 1922 and actively pursued by the late P. R. White in 1931 coeval with work in France. Interest has

centered in factors required for growth and differentiation. Whole organ as well as single cell cultures are maintained. Complete development of the plant has been achieved. The culture techniques are now widely used with possibly ten centers of active research (100).

Chambers designed for growth of plants under controlled conditions were developed at the Boyce Thompson Institute in the 1930 decade (1936, J. M. Arthur) and the U. S. Department of Agriculture, Beltsville (1937, Borthwick and Parker). Engineering development of refrigeration machinery and lighting equipment, as well as an understanding of light requirements for plant growth have facilitated construction of units giving a wide combination of conditions (150). The phytotron at the California Institute of Technology was the prototype for this development. The original phytotron was chiefly devoted to studies of thermo- and photoperiodic requirements of plants. Three other installations fashioned after the Pasadena installation have now been constructed in the United States—at the Universities of Wisconsin, Duke, and North Carolina State. Small cabinet type units are commercially available and are widely used for growth of plants under controlled sets of conditions.

Interest in three aspects of light action on plants is intense in the United States. These aspects are photosynthesis (250), photoperiodism (150), and phototropism (50). Details of the pathway of carbon in photosynthesis following the pattern set by Calvin and his associates continues to attract considerable attention. A breakthrough concerning light action in photosynthesis was accomplished by the late Robert Emerson in 1958. He found that photosynthesis under red radiation is enhanced by far-red radiation. This led to the recognition of a two quantum requirement for photosynthesis. Work in England on cytochromes associated with plastids has led to great interest in cytochromes and quinones as components in the electron transport chain involved in photosynthesis. The trapping center for electrons in system 1 has been recognized as a pigment absorbing in the region of 700 nm. Generation of ATP both by cyclic and noncyclic phosphorylation is under intensive study. Possible involvement of H^+ expulsion as a step in ATP formation, as proposed in England, is being examined. Some attention is turning to the manner in which electron flow, induced by the two quantum absorption, is coordinated.

Measurement of action spectra controlling photoperiodic displays led to recognition (1952) and isolation (1964) of the chromoprotein phytochrome as a controlling factor in light action on growth and differentiation. Etiolation and light requirements for seed germination were also found to be controlled by phytochrome. The part played by rhythmic phenomena in flowering and leaf movement is studied. Short cyclic changes (minutes) in metabolic intermediates have been observed in yeast cells and cell free systems. Studies on details of the

flowering process, which are rhythmic in part, have chiefly been restricted to aspects of morphology.

The most extensive work on phototropism was in the 1930 decade. Failure in finding the nature of auxin action and identifying the photostimulus involved, however, has resulted in a quieting of the subject. Active centers of recent work are Harvard University, California Institute of Technology, and the Smithsonian Institution.

Perhaps the greatest ground-swell now current in plant physiology in the United States is in the area of molecular biology as it functions in development. Most of the work is based on bacteria, but use of higher plants is increasing. An aspect of this work is the possible function of basic proteins (histones) in suppression of gene action. Attention is also turned to the nature of messenger RNA and its functioning in germination as well as the degree of genetic homology between genera as shown by DNA binding.

The American Society of Plant Physiologists is the chief professional organization for the subject in the United States. It was founded in 1924 and shortly thereafter started publication of *Plant Physiology*. This Journal in 1967 contained 1800 pages in the 12 numbers. Membership in the Society is now about 1700 of whom 370 are foreign including 17 corresponding members (Honorary). Research results are also published in the several botanical journals and biochemical ones as well as in *Physiologia Plantarum* and *Planta*. The *Annual Review of Plant Physiology*, of which Vol. 1 appeared in 1950, serves as an international review with more than a third of the reviews from outside the United States. Current volumes contain about 20 reviews of about 25 pages each. Plant physiology, as a subject for research and instruction is presented in possibly 150 universities as part of the curriculum in botany or biology.

The XI International Botanical Congress permits us to assess many of the rapid advances throughout the world during the 5 years since the X Congress. Plant Physiologists in the United States welcome those of other countries to the Congress. Let us hope that some viable seeds for the future will be cast here.

PHYCOLOGY

WILLIAM RANDOLPH TAYLOR

(*University of Michigan*)

The history of algal studies in the United States is, compared with that of terrestrial botany, one of slight beginnings and limping progress even through the 1800's, but which toward the turn of the century exhibits more active research and publication. During the last fifty years our reference materials have reached a level equivalent with that of any other country. In many ways we can now help people from abroad as they, with their classical collections, have long helped us.

As one might expect, the early records of our algae appeared in European books and journals from specimens sent back by travelers or naturalists, and these not abroad primarily to collect algae. Linnaeus in *Species Plantarum* (1753) referred to a few of these species, and mentions Gronovius, *Flora Virginica* (Pt. II, 1743) on p. 1163. Dawson Turner illustrated some American forms even in the first volume of his beautiful *Fuci . . . historia* (1808). Slight and faulty though the current knowledge of American algae was, they were early included in our general floristic manuals, such as Rich's *Synopsis* (1814) where nine genera are briefly described, and Amos Eaton's *Manual of Botany* (1822), but were later more wisely omitted. The most extensive early accounts were those by De la Pyalie of the marine algae of Newfoundland, St. Pierre and Micquelon (1824, 1829) not within our borders, but pertinent to our flora; and Postels and Ruprecht's *Illustrationes Algarum* (1840) contained the giant west-American kelps. *Pterygophora californica, Dictyoneurum californicum,* and the unique *Postelsia palmaeformis* all appear on the magnificent plates of Ruprecht's "Neue oder unvollstandig bekannte Pflanzen aud dem nordlichen Theile des Stillen Oceans" (1852). A few other publications referred to our species, particularly in local lists, one of the more curious being Durant's *Algae and Corallines of . . . New York* (1850), bound with representative specimens. Professor Harvey came from Dublin in 1849 on a lecture tour which took him from New England to Florida. He expressed himself as feeling that he had collected the marine flora rather fully on his trip, but later events have shown the field work must have been far from thorough, and the West Coast flora, for which he depended on correspondents, is hardly represented in his books. He effected arrangements with amateurs on both coasts who supplemented his own collecting substantially, and his three volumes of the *Nereis*

Boreali-Americana (1852-58), while far below the standards of *Phycologia Britannica* (1846-51) still gave us a start with our marine catalogue. Here serious marine studies halted for a time. The period had arrived when ladies at leisure collected seaweeds and pressed them in delicate patterns as vacation souvenirs: some went further and wanted to know the botanical names, and thus specimens got into the hands of experts: their incidental findings are recognized in the publications of Harvey, Farlow and many others, but there was little direct professional work done on marine algae.

Harvey's *Nereis*, sound though it was, was not complete enough to facilitate phytotaxonomic research, and so marine studies were slow to develop. Like freshwater studies they began in the East. Ashmead began to record New Jersey plants as early as 1854, thus contemporary with Harvey, and extended his interests to the Arctic in dealing with Hayes' Smiths Sound Expedition material. The most substantial stimulus now came from Farlow, whose chief fame has come from his work with fungi. He began with compiled lists of New England algae (1873), extended his range to the West Coast (1876) and the Arctic (1879), but worked most productively for his *Marine Algae of New England* (1881-82), which remained our best regional manual for over 50 years. Researches in the southern states did not produce books, but did contribute small collections to northern and European students, such as those of Gibbes and Tuomey from South Carolina, of Blodgett from Florida and Alabama, culminating in the rich and curious collection assembled by Mrs. A. H. Curtiss, chiefly in Florida but including exchange material from elsewhere. East Coast studies now centered about the expert amateur of freshwater and marine algae, the business accountant Collins of Malden and Cape Cod, who carried on Farlow's algal tradition. With the aging amateur Holden, and the younger professional Setchell, Collins started the third of our general American algal exsiccatae, the *Phycotheca Boreali-Americana* (1895-1915), by far the largest ever assembled: it contained 50 fascicles of both freshwater and marine algae. While this included specimens from the West Coast, Alaska and Canada, Collins published little on western algae, but he did interest himself in the tropical flora, with a catalogue of Jamaican algae, and associated with the amateur the Rev. A. B. Hervey, an excellent account of those of Bermuda. His major publication covered all North American green algae exclusive of desmids and charophytes (1909, 1912, 1918), and he projected, but never finished, a major account of New England marine algae. While Collins dominated the field in the Boston area, Howe, then curator at the New York Botanical Garden, moved from his original field of hepatics into marine phycology, contributing substantially to the *Phycotheca* but, since the botanists of the New York area then adhered to the now obsolete American Code of nomenclature, publishing little

with Collins, who worked within the framework of the International Code. Collins' valuable collections eventually came to the New York Botanical Garden. Indeed, Howe published little on mainland algae, but substantially and very critically on Bermudian and Bahaman marine algae. He laid the groundwork for a major publication on Puerto Rican algae, although he did not live to see it to the press.

As neither of these men were in academic situations their influence was slow to penetrate botanical teaching. Here the distinguished personality was I. F. Lewis, Miller Professor at the University of Virginia, but in charge of marine botany at Woods Hole, Massachusetts, for many years where, by great personal charm and wide knowledge, he greatly advanced appreciation of these plants, and spread the fame of the laboratory. Unfortunately, during his tenure, as that of the present writer, the market for phycologists was poor, and few of his Woods Hole students had the chance to publish extensively on algae.

Meanwhile, Setchell had moved from New Haven to the University of California, Berkeley, as Professor of Botany, where he developed our strongest college base for marine botanical research. Many papers came from his laboratories, his own work climaxed by the three (of four planned) volumes on West Coast marine algae, work in which Gardner collaborated. Probably Setchell's most active student was the late Yale Dawson, who published voluminously on marine algae of many widely separated Pacific areas. Since Setchell's death Papenfuss has carried on the traditions of the laboratory, and his students have published on West Coast material while he himself has specialized on the South African flora.

Another center for algal research in the West was established when Gilbert M. Smith moved from University of Wisconsin to Stanford University, where he shifted his interest to experimental studies and to marine algae, producing the excellently illustrated *Monterey Marine Algae* (1944), a work much more widely useful that the title would suggest, and which has been supplemented by Hollenberg and Abbott (1966).

The distribution studies of Pacific Coast marine algae began with Menzies in 1787-88 and 1791-95, which contributed to publications by Dawson Turner (1808-19) and others, supplemented later by those of Saunders (1901) and more completely by Setchell and Gardner (1903). Their more general works have already been mentioned, and for the groups covered they give us our treatment of all the United States West Coast when supplemented by Smith's Monterey book and detailed western lists. A comprehensive catalog for the State of Hawaii is still in the making. On the East Coast only a few local lists existed until Farlow's publication (1881) appeared, dealing with New England, that of Hoyt with the Carolinas (1920), and of the present writer with Florida (1928), but the writer's two volumes in their

revisions (1937, 1957, 1962 and 1960) now cover the East Coast from the Arctic to the tropics, subject to additions which have appeared since, particularly those of Wilce (1959 *et seq.*) which greatly amplify the arctic records.

Studies of freshwater algae took a rather different course. Since freshwater samples made poor vacation souvenirs for travelers, relatively little such material reached Europe in the 18th and early 19th centuries. In this country interest developed through the proliferation of clubs of microscopists, persons who were interested in anything which could be observed with a microscope. They bridged the gap between the professional biologist and the general public, in groups or as individuals. These clubs died with the advent of cheap and convenient cameras and roll film, camera clubs taking their place as diversions. Microscopes had long been in use by the few in Europe: now, with commercial production the many could, and did, venture to enquire into the world of minute things. Lenses were not always good, but those who could afford it competed for the best custommade instruments. These lenses had been made achromatic in a simple form about 1811 and assembled into compound objectives of higher power by 1829. Knowledge of the prevalent defects and how to offset them was common, and a high premium was placed on personal patience and technical skill. Nowhere was greater skill shown than in the study of diatoms, which by reason of their durable siliceous walls lent themselves to amateur study. Competition was keen to show the maximum expertness in demonstrating the resolving power of their achromatic lenses: proud was the microscopist who could resolve the dot pattern in the striae of *Amphipleura pellucida*. Rare indeed is the modern botanist who can do this, even with the best of modern apochromatic optical equipment. The development of the electron microscope has made this a relatively easy matter today using an altogether different technique.

The very real technical skill established through this diatom fad was naturally adapted to observations on other algae, but for a long time few serious floristic or taxonomic studies resulted. Quite reasonably diatoms led the way. Apparently the first serious student of these plants was Professor J. W. Bailey at West Point Military Academy, who, from 1839 for some 30 years, published short notes on freshwater and fossil algae, particularly diatoms. Bailey collaborated with Prof. Harvey in the account of the algae collected on the Wilkes Expedition, illustrated with 9 plates (1874). The climax in publication of the period was the volume on freshwater algae by H. C. Wood (1873) first Professor of Botany, later on Materia Medica, at the University of Pennsylvania, better known for his development of the U. S. Dispensatory. With the excellent library facilities of Philadelphia at his disposal he based his work on Rabenhorst's *Flora Europaearum Algarum,* and

recognized a considerable proportion of the genera in our flora. Even zoologists then recognized algae, and they appear prominently and recognizably in the superb volume on rhizopods (1879) by Joseph Leidy, though only as ingested!

There was a dearth of major works for over a decade thereafter, but interest in microscopy (and at least incidentally in freshwater algae) involving both natural history amateurs and professionals brought forth a number of journals in the last third of the 19th century, for example the *American Quarterly Microscopical Journal* (1878-79), superseded by the *American Monthly Microscopical Journal* (1880-1902), and *The Microscope* (1881-79). Among the earliest and perhaps the shortest-lived was *The Lens* (1872-73), but even in its brief life several algal articles appeared in it. By no means the least of these, the *Transactions of the American Microscopical Society* (1879—) has persisted, becoming more strictly biological and technical through the years, consistently including algal papers. Since the organization of the Phycological Society of America in 1946 and establishment of its journal in 1965 specialists in the field have had a common center of communication.

The Rev. Francis Wolle, Principal of the Moravian Seminary for Young Ladies at Bethlehem, Pennsylvania, with botanical interests ranging from phanerogams through hepatics to algae, in the last quarter of the century began publishing short notes and lists of freshwater species. With extraordinary diligence he produced four volumes (one revised later) covering the desmids (1884, 1892), the freshwater algae in general (1887), and the diatoms (1891), copiously illustrating them himself. Wolle's books, now obsolete for general use, were very popular. Following his death there came another period of little advance.

One must now turn to the Middle West. In connection with the limnological interests so strong at Wisconsin, Gilbert M. Smith made for himself an enviable reputation in freshwater research, culminating in the two well-illustrated volumes on lake plankton (1920, 1924), as well as relatively minor studies, before moving to California. Other regional studies have appeared, but Prescott's on the non-planktonic species of the Great Lakes area (1951) is the most comprehensive. In fact, broad freshwater algal studies have not, in general, received as scholarly attention in America as have the marine species. Exceptions may be made for special groups. For instances, Transeau monographed the Zygnemataceae (1951), Tiffany the Oedogoniaceae (1930), and a succession of workers have published on the Myxophyceae, culminating in the current papers by Drouet and Daily. There has been a notable tradition of attention to the charophytes from the time of the physician T. F. Allen and C. B. Robinson who attempted to list the American forms, to R. D. Wood who has characterized the species on a world basis (1965).

The desmids, too, have received close attention, Wolle having devoted a volume to them (1884) and L. N. Johnson beginning publishing on them in 1894, and Cushman in 1903, though his primary interest was in foraminifera. Gilbert M. Smith illustrated the planktonic forms (1924) and the present writer many of the species found in Newfoundland (1934, 1935), but the major work of recent years has been done, often in collaboration, by Prescott (from 1934), by Hannah Croasdale (from 1935) and by the engineer amateur Arthur M. Scott (from 1950). If we can say that diatoms among freshwater algae earliest received detailed attention, we can also say that interest has continued, chiefly by a few specialists, the secondary school administrator Boyer presenting the first general American flora (1927) since Wolle, while Albert Mann treated of more distant fields (1907, 1925). Ruth Patrick and Reimer are currently monographing the mainland United States species (1966—).

In short, we can say that in the United States we have depended very largely on the great European freshwater floras, such as the series edited by Rabenhorst and Pascher, and have developed for ourselves more specialized, less comprehensive works. Freshwater algae are so often considered cosmopolitan that this is understandable. At the generic level we have Gilbert M. Smith's book (1933, revised in 1950), but that is limited in usefulness. Perhaps as a result our freshwater studies have largely tended to become ecological-limnological, where identification is commonly uncritical.

In summary we may say that the marine algae of our mainland coasts are at least as well documented and their ranges established as those of western Europe. The amount of research carried on by United States botanists on the algae of distant lands, the Arctic, Antarctic, South America, southern Africa, and Pacific islands, that is, almost everywhere except Europe and northeastern Asia, is in notable contrast to Europe-based studies which are generally carried on close to home. These cannot appear adequately in this limited review. On the other hand only our Middle West area freshwater studies approach European standards and completeness. Seldom have American students concerned themselves with freshwater materials from other continents or distant islands.

Life-history and morphological studies of marine algae have largely flourished on the West Coast, particularly at Berkeley and are as yet very incomplete for American species. Freshwater algal life-history studies are not so geographically limited, and the establishment of a very large culture and experimental laboratory by Starr at Indiana University has encouraged such studies, but the field has barely been touched. As to algal cytology there are few important American papers. Cleland has made one of our few meticulous studies on marine algal life history, that of *Nemalion* (1919). In electron microscopy there has been but a beginning, though some nice observations have been made

in California, such as that by Norma Lang (1965) and some diatom studies by Stoermer *et al.* (1964, 1965) but as yet there is no significant trend in this very promising field.

As to ecology, some American work reaches European standards. Certainly the strongest freshwater research center was that which Birge and Juday established in Wisconsin in the early years of this century. Limnological in its approach, it did constrain algal observations into methods more precise than the old comparative species lists involved, and this influenced other laboratories, such as that at Michigan under Paul S. Welch. Studies in flowing waters, too, shared in the advances, the older Kofoid study of the Illinois River plankton (1903, 1908) representing the beginnings of modern methods on a large scale, while the paper by J. L. Blum (1957) on the little Saline River is a later one on a small scale.

Marine ecology has reached a greater sophistication for plankton than for attached forms, and this work has centered at such major marine stations as Scripps Institution of Oceanography at La Jolla, California, and Woods Hole Oceanographic Institution in Massachusetts. The results have been such handsome papers as those by H. B. Bigelow with Lois Lillick and Mary Sears (1940) on the plankton of the Gulf of Maine, and the work has been continued at Woods Hole by Ryther and his associates. Although brackish water studies are more practicable, lamentably little work has been done. At Woods Hole a series of studies on a local brackish "pond" have been detailed, the physical and chemical conditions reported by specialists, the phytoplankton by Hulburt (1956, 1957) and attached algal cycles by Conover (1958). For open coasts descriptive ecology still prevails, and the techniques for relating floras to quantitative physical and chemical factors in the surf zone and below it are still to be perfected.

The economics of algae have received a great deal of speculative appraisal and publicity in this country, but few precision studies. Perhaps the earliest major studies were those related to the use of marine algae, particularly kelps, as fertilizer and a source of potash, and initiated by Turrentine (1912), and under government auspices by Cameron (1914, 1915). Out of the studies of algal physiology, biochemistry and ecology may come something more reliable for detail than what is at present evident, including human sustenance in outer space (*cf.* Burlew, 1953). On a more prosaic level, for instance, Stoloff worked on the chemistry and economics of *Chondrus,* carrageen, for years (*ca.* 1950-57), but probably even more has been done in special laboratories in Nova Scotia and Scotland. Likewise, the economics of the kelp derivative alginic acid has had attention on the West Coast as well as at these same foreign laboratories. There has been very real progress in applying algal studies to utilization of sewage waste with purification and valuable chemical by-products in view, notable studies

being those of Gotaas (1951, 1954), and from the standpoint of the algae as organisms summarized by Silva and Papenfuss (1953).

Physiological studies of algae, including their biochemical precurser studies, have become increasingly numerous, increasingly complex, to the extent that the alga is becoming quite lost to view and is too often recognized only by a culture number. Among the early, primitive observations bordering on the ecological is that of Osterhout (1906) on algae passing rapidly from fresh to salt water with change of tide or by movement of ships on which they grew, but his laboratory shifted in interest to algal respiration and then to permeability, particularly in such large coenocytes as *Valonia*. Studies now seem to have shifted to such matters as subsidiary pigment groups and their chemistry. Fortunately this whole field has been brought together in a manner far beyond the scope of this account under the editorship of Lewin (1962) relating the American to foreign progress in the subject, for this and others see "Selected Readings" references elsewhere in this volume.

PLANT PATHOLOGY AND MYCOLOGY

Kenneth F. Baker

(*University of California, Berkeley*)

Primitive man's belief that unhealthiness in his crops had supernatural causes was questioned only after microorganisms were revealed by the microscope in 1665-84. These organisms were thought to be products of diseased tissues (autogenesis) until it was shown (1861) that they arose only from other identical entities (biogenesis). Some then claimed that microorganisms could attack only plants severely injured by the physical environment (predisposition). Only well after the development of effective pure-culture techniques (1881) and demonstration of the production of plant disease by microorganisms (1807-53) was this restrictive view abandoned.

Knowledge of fungi received a firm foundation from many taxonomic investigations and life-history studies of plant pathogens. Although diseases induced by bacteria, algae, viruses, and nematodes were described by 1880, the causal involvement of these agents was still demonstrated.

The Irish Famine of 1845-48, from a disastrous epidemic of potato late blight, stimulated interest in plant diseases. The prevalent concept of disease control, to destroy pathogens *in situ* by disinfectants, was exemplified by Lister's antiseptic surgery, the use of sprays (lime sulfur) on plants, and chemical treatments for soils. External and internal seed transmission of fungi had been demonstrated by 1882.

The first textbook of plant pathology in relation to agricultural practice appeared in 1858 (Kühn), and the first textbook on mycology in relation to plant pathology in 1884 (DeBary). Sorauer's *Handbuch der Pflanzenkrankheiten,* begun in 1874, has continued to the present.

The last four decades of the nineteenth century thus saw the establishment of mycology and plant pathology through epochal advances in Europe. The U. S. Department of Agriculture was established, and the Morrill Land-Grant College Act passed in 1862, during the Civil War. The first State Agricultural Experiment Stations were thus created (California and Connecticut, 1875), and others rapidly followed. Establishment of these institutions, plus the fact that this country was expanding its frontiers (13 of the 24 states west of the Mississippi River were added after 1875) provided incentive for botanical and agricultural investigations. By 1900 there were 60 Agricultural Experiment Stations.

Plant pathology was first taught in a botany course at the University of Illinois (1873), and as a separate subject two years later at Harvard University. Mycology, begun with the fungus collections of von Schweinitz (1812-32), was first taught in Harvard University about 1875. J. C. Arthur, C. E. Bessey, Farlow, Galloway, B. D. Halsted, R. A. Harper, L. R. Jones, E. F. Smith, Thaxter, and Whetzel were important in shaping early American mycology and plant pathology.

The Section of Mycology, established (1885) under Lamson-Scribner in the U. S. Department of Agriculture, began federal phytopathology, now continued in the Agricultural Research Service. Among early departments of plant pathology were those founded in: Willie Commelin Scholten, Netherlands (1895) by J. Ritzema-Bos; University of California, Berkeley (1903) by R. E. Smith; Royal Danish School of Agriculture (1904) by Rostrup (at age 73!); Canada Department of Agriculture (1906) by Güssow; Cornell University, New York (1907) by Whetzel; University of Minnesota (1907) by E. M. Freeman; University of Wisconsin (1909) by L. R. Jones.

Enthusiasm for study of the new microworld led to neglect of the relationship of the environment to plant diseases; a more balanced ecological viewpoint developed after about 1909. Since 1900 the U. S. has become increasingly important in mycology, and has taken the lead in plant pathology.

The first American periodical in these fields, the *Journal of Mycology* (1885-1908), was succeeded by *Mycologia* (1909) and *Phytopathology* (1910). The American Phytopathological Society, organized in 1908 with 130 members, today has 2440. The Mycological Society of America, founded in 1933, now has 1200 members. Early American phytopathological books were by Scribner (1890), Freeman (1905), E. F. Smith (1905-14), Duggar (1909), Stevens and Hall (1910), Selby (1910), and R. E. Smith (1911).

Continuing major epidemics of plant diseases throughout the world have emphasized the importance of phytopathology. The powdery mildew (1850-52) and downy mildew (1880-85) of grape in Europe, and the coffee leaf rust in Ceylon (1870-95) have been matched in the U. S. by peach yellows (virus) (1806-1936), fire blight (bacteria) of pear (1817-   ), Pierce's disease (virus) of grape (1884-1900, 1935-41), curly top (virus) of sugar beet (1898-   ), citrus bacterial canker (1905-45; successfully eradicated), white pine blister rust (eastern U. S., 1906-; western U. S., 1910-), chestnut blight (fungus) (1906-), phloem necrosis (virus) of elm (1918-), Dutch elm disease (fungus) (1930-), Stewart's bacterial wilt of corn (1931-34), stem rust of wheat (1935-37), after control by resistant varieties for many years), tristeza (virus) of citrus (1939-), oak wilt (fungus) (1943-), and pear decline (virus?) (1951-). These dramatic aspects of disease are probably less important than the annual crop losses; in 1951-60,

for example, annual disease losses in the U. S. averaged $3,691,914,000.

Early plant pathology in the U. S. emphasized checking disease outbreaks, but this soon gave way to preventive treatments based on knowledge of pathogen biology. The *art* of controlling plant disease thus contained, from the beginning, rudiments of the *science* of phytopathology. Studies on the physiology, biochemistry, and ultrastructure of host-parasite interaction in disease inception and development increased after the late 1930s. With available grant funds for such basic research ₊after 1950, and less urgent need for new disease controls, applied plant pathology decreased. This trend has caused some concern that the field may largely become a study of physiology of the abnormal plant.

Fungi, Bacteria, and Nematodes:

The ability of bacteria to induce plant disease was demonstrated by Burrill and Arthur (1879-85) and transmission of bacteria by bees was soon shown by Waite (1889-91), the first demonstrated insect transmission of any plant pathogen. Although phytobacteriology expanded rapidly in America for 40 years under the leadership of E. F. Smith, it then declined somewhat until after 1960.

The autoclave was developed (France), and Koch's Postulates (Germany) for proving pathogenicity stated in 1884. The Postulates came, by 1931, to represent a stricture to plant-disease studies, because they ignored accompanying microorganisms, and did not apply to obligate parasites. Saccardo's *Sylloge Fungorum* (1882-1931) and Engler and Prantl's *Die Natürlichen Pflanzenfamilien* (1897-1907) were major taxonomic aids. Klebs (1900) elucidated the conditions necessary for fungus sporulation. It was found (1967) in Australia that some obligate parasites (rusts) could be grown in pure culture.

The demonstrations (1894-1903) in Europe of physiological races of fungi revealed difficulties in developing resistance to these pathogens. The sexual function of the pycnia in rusts was proved in Canada (1927), and it was discovered (U. S., 1930) that hybridization on barberry produced new races of stem rusts of wheat. Hansen and Smith (1932) showed that Burgeff's heterocaryosis (1914-15) provided another mechanism by which new strains were developed. Flor (1956) postulated that a gene for pathogenicity in the fungus was matched by a host gene for resistance.

Materials inhibitory to fungus pathogens in the normal host were first shown (1929) with the isolation of protocatechuic acid from onion scales. Phytoalexins (inhibitory materials produced by the host after stimulation by a pathogen) were postulated in 1940 (Germany). The first such material isolated was pisatin (Australia, 1960-61). Active agents produced by *Fusarium moniliforme,* which cause the well-known "bakanae effect" on rice, were found (Japan, 1935) to be

the gibberellins. These substances were shown (U. S., 1956) also to be hormones in higher plants.

It was shown in Canada (1950) that germination of fungus spores was inhibited in field soil, and a few years later this phenomenon (fungistasis) was further defined in Britain. It was demonstrated (U. S., 1964) that a microorganism-induced nutrient deficiency might produce this effect. Dormancy factors are also known to exist in spores.

Electron-microscope studies of the ultrastructure of fungi, bacteria, and viruses became common and of good quality in the 1960s.

Although descriptive nematology was well started before 1900, the field of phytonematology largely developed after about 1935. Although Cobb (1907-32) had provided early U. S. leadership in this field, the first university course in nematology was not taught until 1947 (California). Biological information, methods of nematode feeding (1936-37), details of physiological races (1944) known for 20 years to exist, and pure-culture studies (1914-55), were pursued. Nematology developed in the U. S. as part of plant pathology, in contrast to the situation in Britain and Europe. The Society of Nematologists was formed in 1961.

VIRUSES:

While plant-virus diseases were observed in Europe in the sixteenth century, and their transmission by grafting was known at least from 1714, their filterability and mechanical transmissibility were not demonstrated until 1885-98. The study of plant viruses has become a major area of plant-pathology research in the last 50 years. Tobacco mosaic virus has provided exceptional experimental material, and studies on its structure gave great impetus to research in biochemistry and molecular biology.

Viruses were shown in the U. S. to produce no symptoms in some hosts (1918) or under some environmental conditions (1922), and were found in Canada to produce intensified symptoms (synergism) when two viruses were present (1925).

Virus strains were demonstrated (U. S., 1925), and the relationship between them could be approximated by intensity of immunological and serological reactions. Quantitative studies were introduced into virology by Holmes' local-lesion method (U. S., 1928), and were later supplemented by the ultracentrifuge (U. S., 1936), electrophoresis (U. S., 1930), electron microscope (Germany, 1939), and by density-gradient centrifugation (U. S., 1953).

Attempts to purify viruses became intense after 1930, and several laboratories were close to success when Stanley (U. S., 1935) achieved it. Virus infectivity was found (U. S., 1956) in the RNA rather than in the protein fraction of the particle.

Studies on arthropod vectors have shown leafhoppers (U. S., 1902-

10), aphids (U. S., 1912), chewing insects (U. S., 1924), thrips (Australia, U. S., 1930), white flies (Africa, 1930), mealybugs (Africa, 1945), and mites (Canada, 1953) to spread plant viruses. It was shown (Africa, 1932) that some leafhopper strains may transmit a virus, and others not, that some insects can only acquire a virus while in the larval stage (Australia, U. S., 1931), that viruses may increase in an insect as well as in the plant (U. S., 1941), and may even be passed through the eggs for several insect generations (Japan, 1933). Some viruses may persist in the insect during its life, others for only a short period. Various viruses in plants may largely be confined to the parenchyma tissue, the phloem, or the xylem. Nematodes (U. S., 1958) and fungi (U. S., 1962) also act as vectors. Seed transmission of viruses was shown in the U. S. to be internal (1919) and external (1937); fortunately, neither type is common. Cultural practices (*e.g.*, transplanting, pruning, grafting) provide very efficient means of spreading viruses in crops.

Latin binomials, used for viruses during 1927-56 but replaced for a time by common symptom names, are reappearing.

DISEASE CONTROL:

Although integrated control procedures against a plant disease were suggested by Jensen in 1882, it was 70 years before this principle was widely applied. H. M. Ward (1882) showed that fungicides usually must be applied to the host prior to infection (preventive treatment). Jensen also demonstrated the use of thermotherapy against internal propagule-borne fungus pathogens (1882, 1888). This was soon extended (Kobus, 1889) to viruses, and was clearly demonstrated in 1923 and 1934. Widespread chemical (CS_2) treatment of soil was practiced after 1869, but declined after 1900. These studies were made in Europe, Ceylon, and Indonesia.

Steam was used for commercial soil treatment (U. S.) in 1893; the cooler aerated steam was adopted after 1960. Large-scale chemical soil treatment recommenced after 1934-35 (Hawaii) with chloropicrin against nematodes and fungi, but cost restricted its use. A cheaper material, D-D, introduced (1943) for nematode control, proved so profitable that growers were stimulated to try costlier materials, which also proved economic. These treatments all attempted to eliminate the pathogen at source (*i.e.*, in host plants and soil) before they reached the suscept, and thus resembled in principle aseptic surgery, which came into medical practice about 1890.

The dominant method of disease control, destruction of the pathogen at the infection site, received strong reinforcement from the accidental discovery of Bordeaux mixture in France (1885). Sprays were applied by high-pressure equipment in the U. S. after 1894. Fungicidal dusts were used in the U. S. after 1907, and were applied by airplanes after 1921. When it became obvious after 1917 that time of application

was as important as quantity, meteorological spray-timing services became common. Prediction of epidemics also developed after 1920, greatly assisting preventive treatment. The destruction of the carry-over stage of the pathogen in or on the host by eradicatory dormant sprays, developed in California (1880-85) against peach leaf curl, was extended to other diseases after 1924.

The first patented fungicides, organic mercury compounds for seed treatment, appeared in Germany in 1914. The alkyldithiocarbamates were introduced (U. S.) in 1934. Intensive industry efforts in the last 20 years have produced a remarkable number of new organic fungicides of high specificity and low plant and mammalian toxicity. Thus ended the reign of lime-sulfur and Bordeaux sprays.

Quarantines became increasingly important in the U. S. after they were first established in 1912. The elimination of virus carry-over by a host-free period has been used in California since 1933. Seed of some crops were produced in areas where the environment precluded development of a pathogen. Indexing methods were devised to select propagules free from a pathogen (*e.g.*, certified seed potatoes, bean and lettuce seed). A culturing technique (1943) to select material free from *Verticillium* was extensively used. Small tip cuttings, growing points, and (after 1954) minute meristem tips, alone or combined with thermotherapy, were widely used to eliminate virus from infected varieties of many crops. Prolonged thermotherapy also produced many types of virus-free plants. Interest in pathogen-free propagules was keen in the 1960s.

Studies of biological control of root pathogens, begun in 1921, were pursued vigorously in the 1960s. While the books of Rachel Carson (1962) and Rudd (1964) had less impact on fungicide than on insecticide use, they provided impetus to non-chemical control measures. Phytotoxic atmospheric pollutants, long a localized problem associated with industry, became increasingly serious in many areas after 1950; a new research field thus appeared.

Beginning with Cobb and Farrar in Australia (1890-98) on wheat stem rust, Orton in the U. S. (1899-1909) on fusarium wilts of several crops, Bolley on fusarium wilt of flax in the U. S. (1903), and Biffen in England (1905-12) on wheat yellow rust, there has been great success in developing resistance to many plant pathogens. Vavilov (1935) started using wild species from the native homes of crops as sources of resistant germ plasm. Monogenic resistance was sought by breeders for a time, but its specificity in the face of rapidly appearing new pathogen strains brought multigenic (horizontal) resistance into favor in the 1960s.

Epilogue:

The 8000 years that man has grown crops have seen extensive selection for useful characters, which made some crops (*e.g.,* corn) entirely

dependent on him for survival, and generally lowered their resistance to disease. Monoculture over large continuous areas led to epidemics. Modification of the environment by such practices as overhead sprinkling, increased fertilizer application, and growing plants in marginal climates or seasons, have greatly favored disease, while extending crop range or yield. Although disease resistance or control procedures have thus made possible many cultural improvements and have contributed to greatly increased yield, disease losses today probably represent the same relative percentage of the crop that they did in ancient agriculture. Plant pathology, in this view, enabled man to improve cultural practices and yield, while preventing increase in disease losses. Sudden adoption of modern cultural techniques or varieties in agriculturally underdeveloped countries, without necessary disease studies, may lead to serious epidemics. The present world food crisis will inevitably bring plant pathology into greater prominence.

Mycology and phytopathology have made many important contribuitons to science. Among these are the discovery of penicillin (1928) and other antibiotics, of gibberellin plant hormones, and the stimulation to molecular biology provided by studies on tobacco mosaic virus. Studies on virus movement in plants provided data on rate of translocation in phloem tissue. Knowledge of root exudates largely came from studies on their importance to rhizosphere microorganisms. A new dimension was added to host-parasite interactions by demonstrating fungus-inhibitory phytoalexins formed by the host upon invasion. Many details of plant anatomy resulted from disease studies. Studies on the ultrastructure of the host and bacteria and fungi, and their mechanisms of interaction, were greatly stimulated. Knowledge of insect feeding on plants was increased by virus transmission studies. Genetics and plant breeding received impetus from the necessarily continuous breeding for disease resistance. This also contributed the gene-for-gene hypothesis, improved understanding of the nature of disease resistance and fungus genetics, and stimulated studies on the origin and evolution of cultivated plants. Plant pathology and mycology have thus provided grist for man's mind as well as his digestion.

BRYOLOGY AND LICHENOLOGY

Emanuel D. Rudolph

(*Ohio State University*)

The reader may wonder at the marriage of bryology and lichenology, two fields that on the surface seem quite different. However, the union is not completely illogical because in the United States these two fields had a real connection during their developmental period, and they remain yet connected by the inclusion of lichenologists in the membership of the American Bryological Society and of lichenology in its journal.

The descriptive phase of bryology and lichenology in the United States began as soon as mosses, hepatics, and lichens collected by botanical explorers were sent to experts for identification. It is no surprise that much of the collecting and almost all of the identification of this "Beginning Period" were done by Europeans.

THE BEGINNING PERIOD: Before 1838. The first United States lichen on record is probably *Cladonia sylvatica* (L.) Hoffman, the specimen sent by Banister to England from Virginia between 1678 and 1692, and another by Clayton to Gronovius who included the species in *Flora Virginica* (Pt. II, 1743). Linnaeus referred to *Flora Virginica,* and to Dillenius, *Historia Muscorum* (1741) which included American species. The first moss is more difficult to identify since the genus *Muscus* at that time included almost all cryptogams, but the first notice of bryophytes and lichens in a flora of the United States mainland was in *Flora Virginica*. American bryophytes and lichens were mentioned in Sloane's *Catalogus plantarum Jamaicae* (1696), and in Plumier's *Nova plantarum Americanarum genera* (1703). Floras of the latter part of the 18th and early 19th centuries that included these two groups were Walter's *Flora Caroliniana* (1788), Muhlenberg's "Index flora Lancastriensis" (1793), and Michaux's *Flora boreali Americana* (1803). Muhlenberg had exchanged specimens with Hedwig, and William Baldwin in 1811 sent "a great number of Mosses and Lichens" to Olaf Swartz and Acharius from southeastern United States, and in 1815 to Schwaegrichen. Swartz, Baldwin recorded, returned an answer "confirming" his names. In 1812 Baldwin sent a representative part of his moss collection from southeastern United States to Muhlenberg, and in 1819 enroute to meet the Long Expedition he collected in western Pennsylvania.

Swartz included two New England strays among his Jamaican lichens in his 1811 publication on the American species, and at this period American mosses are to be found in the general writings of Hedwig, and lichens in the writings of Acharius. Muhlenberg's *A Catalogue of the Hitherto Known and Naturalized Plants of North America* (1813, rev. 1818), of which an abbreviated version of the cryptogamic portion is to be found in Eaton's *Manual of Botany* (1817), lists approximately 175 taxa of bryophytes and lichens. Muhlenberg was one of the first botanists in this country to take more than a passing interest· in cryptogams. He sent specimens to such active Europeans as Palisot de Beauvois, Hedwig, Hoffman, and Swartz, in whose writings, as well as in those of Dillenius, Linnaeus, Acharius, W. J. Hooker, Bridel, and Dawson Turner, American bryophytes and lichens had been reported. Soon native botanists were publishing: on bryophytes, von Schweinitz (1821) and L· C. Beck (1829), and on lichens, Halsey (1823). Among local floras of the period which included cryptogamic plant groups are Torrey's *Catalogue of the Plants . . . of the City of New York* (1819), E. Hitchcock's *Catalogues of Animals and Plants of Massachusetts* (1833), and Riddell's *Synopsis of the Flora of the Western States* (1835) that especially covered the Ohio River Valley. The "Beginning Period" can be said to have provided rudimentary information about the bryophytes and lichens of the United States—their names, numbers, and distributions. Much of the critical work was done by Europeans who, except for von Schweinitz and Halsey, were the only specialists for these plant groups.

THE SULLIVANT-TUCKERMAN PERIOD: 1838-88. Notices appeared in the literature on several occasions about a projected cryptogamic flora of North America to be prepared by von Schweinitz and Halsey, and perhaps John Torrey. This flora, as well as the projected moss flora by Beck never materialized. Inclusive floristic treatments of American mosses and lichens were not written until the end of this period, and then only as a result of much more widespread collecting and careful study. The two men primarily responsible for the investigations leading to the publication of these major floras were Sullivant, a businessman of Columbus, Ohio, and Professor Tuckerman of Amherst. Both began their botanical careers studying flowering plants. Sullivant began to concentrate on mosses about 1840 and published his first paper "Contributions to the Bryology and Hepaticology of North America" in *Proceedings of the American Academy of Arts and Sciences* (Vol. 3, 1846). A year earlier his *Musci Alleghanienses* (2 vols.) appeared which included printed labels of moss specimens collected on a trip made with Asa Gray.

From then until his death Sullivant, with patience for detail and scrupulous accuracy, was the leading bryologist in the United States

and was internationally recognized as such. He determined the bryophyte collections of most of the American exploring expeditions, notably those of the Wilkes Expedition on the northwest coast of the United States, the South Pacific, and Antarctica. Sullivant's illustrated treatment of "Musci and Hepaticae" in the second edition of Gray's *Manual of Botany* (1856), together with his greatest work, *Icones Muscorum* (1864) which contained 129 copperplate illustrations of "most of those mosses peculiar to eastern North America which have not been heretofore figured," and its posthumous supplemental volume (1874) established American bryology as comparable with that of Europe.

Edward Tuckerman began his studies of American lichens about 1838, with the later encouragement of Elias Fries whom he had met in Europe, and of Asa Gray. His first papers concerned New England lichens as he had been the first to collect them, but by 1845 he had published "An Enumeration of North American Lichens." A supplement to this described many new species from California and southern United States (1858 and 1859). He wrote the lichen portions of the reports of the American exploring expeditions as Sullivant had for the mosses. *Lichens of California, Oregon, and the Rocky Mountains* (1866) was such a report. For most of this period Tuckerman was the only American lichenologist, or "lichenist" as they were called in those days, but in his final twenty years he was joined by his pupil Henry Willey. Tuckerman, like Sullivant, published on lichens of many parts of the earth, issued sets of identified lichen specimens, and gained international recognition. The first volume of his flora, *Synopsis of North American Lichens* (1882), appeared four years before his death, and the second volume, completed by Willey, two years after his death. Tuckerman's *Genera Lichenum: an Arrangement of North American Lichens* (1872) is regarded as his greatest work. A conservative taxonomist, he was skeptical of Schwendener's hypothesis of the dual nature of lichens and remained until the end a follower of the Friesian system of classification. His pupil Willey was even more conservative and an even more vituperative opponent of the dual hypothesis.

Lesquereux who had completed the supplement of his *Icones Muscorum,* together with T. P. James who had consulted with Sullivant during his first twenty years as a bryologist, brought Sullivant's dream of a continental moss flora to fruition with the publication of *Manual of the Mosses of North America* (1884). Sullivant had several younger colleagues besides Lesquereux and James. There were C. F. Austin, a New Jersey specialist in hepaticology who, like Sullivant, issued several sets of *exsiccatae;* Eugene A. Rau, and Wolle, both of Bethlehem, Pennsylvania. Sullivant was fortunate in having the extensive moss collections taken by Drummond in the southern United

States and Canada, identified and distributed by Hooker. Both Sullivant and Tuckerman benefited by having specimens collected in various parts of the country by numerous correspondents, among them Bolander from California and Nevada. Bolander on his arrival in Columbus, Ohio, at the age of fifteen had come under the influence of Lesquereux, and from this early contact had developed a life-long interest in mosses. Lesquereux wrote in 1869 that Bolander in less than a year in California had collected as many species of mosses as all the other collectors together.

Other collectors sent specimens to Sullivant and Tuckerman:

Beaumont	from Alabama and Florida
Mohr, and **Peters**	Alabama
Josiah Hale	Louisiana
M. A. Curtis	North Carolina and Virginia
Ravenel	**Tennessee, South Carolina, and Texas**
C. C. Frost	Vermont and New Hampshire
Olney, and J. L. Bennett	Rhode Island
E. Michener	New Jersey and Pennsylvania
Peck	New York
T. C. Porter	Pennsylvania
Russell	New England
W. Oakes	Massachusetts
T. G. Lea	Ohio
Elihu Hall	Illinois, Missouri, Kansas, Colorado, and Oregon
Hayden	Nebraska and the Rocky Mountains
Shepherd	Missouri
Sereno Watson	California and Nevada
Charles Wright	California and Texas
J. G. Cooper	California
Fendler	New Mexico

By 1888 the first comprehensive floras of American bryology and lichenology were finished. About 900 species of mosses and 1,200 species of lichens had been described. Under the leadership of Sullivant and Tuckerman who had used European models and the advice of European specialists, America had reached a standard comparable with that of Europe and a firm foundation had been laid.

PERIOD OF TRANSITION: 1888-1947. With improved microscopic lenses and laboratory techniques European cryptogamic botany became

increasingly concerned with comparative morphology and life cycles. Germany led in this approach and her influence was not long in making its mark on the United States. The outlook of the new generation of American cryptogamic botanists is well exemplified by Professor Farlow who, when reminiscing in 1913 about his studies with De Bary at Strasburg, said, "For the last twenty years most young American botanists have thought it necessary to study in Germany to complete their education, but, when I returned in 1874 I was looked upon very much as one would who had returned from a journey in Thibet or Central Africa." Although he did not concentrate on either bryophytes or lichens, Farlow was the leader of the new cryptogamists, and he did write some papers which included these two groups. Others who studied in German laboratories during the early part of this period included the bryologists Campbell of Stanford who studied with Strasburger and Pfeffer, and Evans of Yale, and the lichenologist Dr. Hasse, a physician at Soldiers' Home near Los Angeles. At the same time interest in floristic and monographic studies also increased. The works of some of the older men exemplified this new emphasis: D. C. Eaton's *A Check-list of North American Sphagna* (1893), the arrangement following Warnestorf, and with Fink, *Sphagna boreali-Americana exsiccata* (1896); Underwood's *Descriptive Catalogue of the North American Hepaticae north of Mexico* (1884), and his contribution to the 6th edition of Gray's *Manual* (1890); and C. R. Barnes' "Analytical Key to the Genera of Mosses" (1886). Verdoorn's comprehensive *Manual of Bryology* (1932), and Steere's short review "Bryology" (1955) include bibliographies that record the achievements of this period on an international scale. Bruce Fink, the foremost lichenologist of this period, reviewed the changes he saw taking place before 1904 in his "Two centuries of North American lichenology."

A most important development was the founding in 1898 of *The Bryologist,* for its first two years a part of *Fern Bulletin,* but thereafter an independent publication. The Sullivant Moss Chapter of the Agassiz Association was responsible for its publication, and became an independent chapter in 1899. Grout seems to have initiated the chapter. He was aided during the early years by Mrs. Elizabeth Britton and Carolyn Harris. In 1908 the Sullivant Moss Chapter changed its name to Sullivant Moss Society, and in 1949 to The American Bryological Society. In 1901 lichenology became an activity of the society and has been ever since. Detailed bryological local floras appeared:

California (hepatics) M. A. Howe (1899)
Connecticut (general) A. W. Evans and G. E. Nichols (1908)
Florida (hepatics) H. Kurz and T. M. Little (1933)

Kansas (mosses) Minnie Reed (1894)

Michigan (hepatics) Steere (1940)

New York (mosses) Grout (1916)

Ohio (mosses) Nellie F. Henderson (1927-1931)

Oregon (hepatics) Ethel I. Sanborn (1929)

Pennsylvania (mosses) O. E. Jennings (1913)

Tennessee and North Carolina (general) A. J. Sharp (1939)

Vermont (mosses) Grout (1898)

Washington and Idaho (mosses) George Neville Jones (1930)

Washington and Oregon (general) Lois Clark and T. C. Frye (1928)

West Virginia (hepatics) Nellie Ammons (1940)

West Virginia (mosses) Elizabeth G. Britton (1892)

Wisconsin (mosses) L. S. Cheney and R. Evans (1944)

Checklists of North American bryophytes were issued in 1885, 1925, 1938, and 1940. The last, by Grout, Andrews, and Evans contained 460 hepatics and 1,166 mosses. During this period Grout edited and personally published *Moss Flora of North America north of Mexico* (3 vols., 1926-40), and T. C. Frye and Lois Clark, *Hepaticae of North America* (1937-47). These works incorporated the considerable amount of information contained in monographic and floristic studies that had been published during this period· Several accounts of monographic extent by Andrews, E. B. Bartram, E. G. Britton, A. W. Evans, Fulford, Grout, M. A. Howe, R. S. Williams, and Wynne appeared in some volumes of *North American Flora* (New York Botanical Garden).

Popularization of moss studies begun by Mrs. Britton in her articles in *The Observer* (1894), and by Grout in *Mosses with a Hand-lens* (1900), together with the activities of the Sullivant Moss Society, stimulated serious amateurs to collect and study bryophytes. Other popular works that appeared in the latter part of the period included Elizabeth Dunham's *How to Know the Mosses* (1916, rev. 1956), Bodenberg's of Mt. Rainier National Park (1939), and H. S. Conard's *How to Know the Mosses* (1944; rev. including liverworts, 1956). The rapidly accelerating instructional interest in comparative morphology and its evolutionary implications are evidenced in Campbell's *Structure and Development of the Mosses and Ferns* (1895; ed. 2, 1905), and again in *Evolution of Land Plants* (1940). Both he and Barnes placed strong emphasis upon the morphology of bryophytes. In 1917 C. E. Allen discovered sex chromosomes in hepatics, and continued genetic studies for many years. Toward the end of this period other aspects of bryology, including physiology, genetics and ecology, were initiated

by R. F. Griggs, S. L. Meyer, P. M. Patterson, Margaret Fulford, and their associates, while anomalies in distributional patterns were elaborated by Sharp, Steere, and their associates.

Culberson in "A Guide to the Literature of the Lichen Flora and Vegetation of the United States" (1955) lists such older devotees as Hasse and Calkins; and its newer students Clara Cummings, Fink, G. K. Merrill, Plitt, and Riddle. New detailed floras appeared:

California   Hasse (1895-1915) ; Herre (1906-50)

Illinois   Calkins (1896)

Iowa   Fink (1895)

Maine   Parlin (1939)

Michigan   Joyce Hedrick (1931) ; Hedrick and Lowe (1937) ;
   Steere (1940)

Minnesota   Fink (1896-1910)

West Virginia   Sheldon (1939)

Monographic studies of lichens were made by C. W. Dodge (*Stereocaulon*), A. W. Evans (*Cladonia*), G. T. Johnson (*Trypetheliaceae*), R. H. Howe, Jr. (*Usneaceae*), and Riddle (*Parmeliopsis* and *Stereocaulon*). A major new flora containing descriptions of about 1,500 species written by Fink, but completed after his death by his student, Joyce Hedrick, appeared in 1935. Nevertheless the popularization of lichenology in this period was less substantial than in bryology, perhaps because of the lack of a simple usable guide like *Mosses with a Hand-lens,* and a leader like its author. Articles by Carolyn Harris and F. L. Sargent appeared in early issues of *The Bryologist;* A. Schneider wrote a guide (1898, 1904) ; and Nearing's *The Lichen Book* (1941-47) was appearing, since reprinted (1962), yet amateur lichenologists have never been numerous.

The new interest in morphology and physiology manifested itself in lichenology in the writings of Curtis, G. T. Johnson, G. J. Peirce, Schneider, and Sturgis. The dual nature of lichens was championed by Fink, and soon it became generally acknowledged as a fact. Evans, originally a student of hepatics, became a leader in the use of chemical tests for identification and definition of lichen taxa in his studies of the genus *Cladonia,* the tests brought together by Thomson (1967). They became more important in the works of the latter part of this period.

The "Period of Transition" was one during which detailed information was accumulated on the distribution, structure, and to a lesser extent, the biology of bryophytes and lichens. By the end of 1947 significant new national floras of the mosses, hepatics, and lichens had been published. Some Europeans were still doing substantial work

with American bryophytes and lichens — for example, Degelius, Gottsche, Kindberg, Magnusson, Räsänen, Röll, Stephani, and Warnstorf, but most of the studies were being made by Americans who were contributing to other aspects of their fields as well. The number of students of both groups of plants was increasing by the end of this period, and the two fields were in a healthy state.

THE RECENT PERIOD: 1947— Interest in physiology, morphology, and particularly cytology and genetics, has increased. While no national floras have been completed, new checklists have appeared, such as of mosses by Crum, Steere, and Lewis Anderson in *The Bryologist* (1965), and of lichens by Hale and Culberson in *The Bryologist* (1966). The mosses now number 1,159 species not very different from the number in 1940, but the number of lichens in the latest checklist is 2,558 species, almost double the number in Fink's 1935 flora. Among the more extensive local moss floras are: Pennsylvania: (O. E. Jennings, 1951), Indiana (Winona Welch, 1957), Florida (Ruth Breen, 1963), and Michigan, (Darlington, 1964); hepatic floras of New York, (Schuster, 1949), and Minnesota (Schuster, 1953-58); and lichen floras of Washington (Grace Howard, 1950), western states (Imshaug, 1957), South Dakota and Wyoming (C. M. Wetmore, 1967), Ohio (C. J. Taylor, 1967), and New York (Brodo, 1968). Monographic studies continue to increase, and many can be found in recent issues of *The Bryologist*. The genetics and chromosome behavior of bryophytes has been increasingly studied, and investigation of polyploid species by Steere and his associates has provided new insights into their speciation. Cytology of bryophytes, but to a lesser extent of lichens, now includes electron microscopy as well as light microscopy. The more common isolation of bryophytes and lichen symbionts in bacteria-free culture has made them useful organisms for physiological experimentation. A number of American laboratories have active programs with these types of cultures.

In lichenology the nature of the symbiotic relationship has been greatly advanced by the work of Ahmadjian and his associates, a good summary of which is his recent book, *The Lichen Symbiosis* (1967). Studies of whole-thallus lichen physiology and ecology are becoming more common in this country. Llano's reports on the economic uses of lichens (1948 and 1956) and Thieret's on mosses (1956) foreshadowed the current interest in these plants as accumulators of radioactive fallout and as contributors to the arctic food web. Lichen taxonomists have continued to show interest in lichen chemistry and in ecological variation. The appearance of two books by Hale, *Lichen Handbook* (1961) and *The Biology of Lichens* (1967), has made recent developments available to other biologists.

The "Present Period" of bryology and lichenology has been characterized by a broader biological approach. A future history will no doubt record an accelerating trend.

PTERIDOLOGY

Rolla Tryon

(*Harvard University*)

The earliest reports on American ferns came from European works such as that of the French botanist Cornut who, in his *Canadensium plantarum* (1635) illustrated two American species. His plates of *Filix baccifera* (= *Cystopteris bulbifera*) and of *Adiantum americanum* (= *Adiantum pedatum*) are very creditable renditions of the species. The limited collections and knowledge of the ferns during this early period is reflected in Linnaeus' *Species plantarum* (1753) which accounted for only 23 species from the area of the United States and Canada. André Michaux in his *Flora Boreali-Americana* in 1803 included 55 species of Pteridophyta, and Frederick Pursh in *Flora Americae septentrionalis* (1813, "1814"), including a larger area, treated 98 species.

By the time of these last works there was a significant number of resident scientists concerned with botanical studies and the first editions of such manuals including ferns as Bigelow's *Florula Bostoniensis* and Eaton's *Manual of Botany for the Northern States* were published from 1814 to 1817. The initial dependence of American botanists upon those abroad was replaced by mutual collaboration as botanical centers with libraries and herbaria became established. This transition was accompanied at times by a certain rivalry, for example, when scientific patriotism was much aroused by the publication of Pursh's *Flora* in London. The early American contributions were devoted almost entirely to general floristic studies. Exceptions were the account of the diminutive *Botrychium simplex* by Edward Hitchcock in 1823, the first new fern described by an American, and the "Synoptical tables of the ferns and mosses of the United States" by Beck in 1829.

The latter half of the 19th century witnessed the real beginnings of American pteridology. The report by Brackenridge in 1854 of the "Filices, Lycopodiaceae and Hydropterides of the United States Exploring Expedition of 1838-1842, under the command of Charles Wilkes," (with the folio atlas published a year later) was the first important publication devoted to the ferns. Brackenridge produced a remarkable piece of work, considering that it dealt with collections made throughout the tropics, and that he had neither special training for the task nor adequate facilities for its completion. This was the first American effort abroad and began a tradition of work in the tropics. The collections of Wright from Cuba and those of Fendler

from Venezuela came, at about the same time, to Asa Gray at Harvard. These formed the subject of the Doctoral Dissertation of Daniel Cady Eaton in 1860. "Filices Wrightianae et Fendlerianae," which was a critical enumeration of nearly 360 species.

Eaton later had a distinguished career at Yale University where he established his position as the first American pteridologist. He prepared several treatments on fern collections from the western United States and Mexico and described many new species. He contributed the account of the Filices in the 5th edition of Gray's *Manual of Botany of the northern United States* in 1867 and of the Pteridophyta in the 6th edition of 1890. Eaton's major contribution was his two large quarto volumes, *The Ferns of North America,* published in fascicles 1877-80, treating 149 species fully illustrated with colored plates. It is not only the first fern flora of the United States (and Canada) but is the most detailed and elaborate ever published. A year later Underwood published a modest treatment, *Our Native Ferns and how to study them,* which was based on Eaton's "Conspectus" in the second volume of *The Ferns of North America.* Underwood's small manual was very popular and went through six editions—the title changed to *Our Native Ferns and their Allies* and the text was expanded in the second and subsequent editions—the last in 1900 treating 279 species of Pteridophyta.

Engelmann had a remarkable botanical (as well as medical) career. He was the preeminent American monographer through his work on such groups as *Cuscuta, Pinus, Vitis,* and the Cactaceae, and in 1882 he published the first American monographic treatment on the Pteridophyta, "A revision of the American species of *Isoëtes.*" In this difficult genus Engelmann distinguished twelve species and fourteen varieties, describing eighteen of these taxa for the first time. Engelmann's interest in Pteridophyta was evident much earlier when he provided translations of papers by Braun and Kunze for publication in the *American Journal of Science* (1848). He added some of his own observations to these studies on *Equisetum, Marsilea,* and ferns on the basis of his own materials. The manuscript of the translation of Kunze's paper, "Notes on some ferns of the United States," evidently passed through many hands for it also contains notes added by Braun, Gray, and Silliman, the editor. The materials of *Isoëtes* in Engelmann's herbarium contain many drawings and notes that illustrate his meticulous methods of study. His specimens are at the Missouri Botanical Garden where his herbarium, combined with that of Bernhardi form the nucleus of the present large collections. Pteridology was well established by 1885. The ferns of the United States were becoming fairly well known and studies in the American tropics had begun. Floristic and monographic studies that had been published set a high standard for future work.

A popular interest in ferns developed rather independently of the technical studies. The Chautauqua Movement for the popularization of knowledge stimulated a broad interest in natural history. This is discussed by M. Curti, *The History of American Thought* (Harper Bros. 1943). The Linnaean Fern Chapter of the Agassiz Association, organized by Clute in 1893, provided a special focus for fern study. The first number of the Chapter's *Linnaean Fern Bulletin* was published that year. This small journal—the first few numbers were of eight to twelve pages, about three by five inches, "price 5 cents"—was the beginning of the *Fern Bulletin* which after twenty volumes became the *American Fern Journal*. The Linnaean Fern Chapter later became the American Fern Society. Clute's role in the establishment of the *Fern Journal* in 1912 is an interesting one. He had been publishing the *Fern Bulletin* privately (selling copies to the Fern Society for its members) but his editorial control and policies were considered so oppressive by some of the members that they agreed to launch a new journal. In spite of Clute's protests that there was a lack of meritorious papers for any new venture, the *American Fern Journal* was an immediate success. The American Fern Society now has about 700 members and its quarterly journal, in its 59th volume, serves an effective role in promoting the interests of all aspects of pteridology.

Popular manuals on ferns have done much to direct attention of the amateur to the study of the group. The most influential of the early ones were Parsons' *How to know the Ferns* (1899), Clute's *Our Ferns in their Haunts* (1901), Waters' *Ferns* (1903), and Tilton's *The Fern Lover's Companion* (1922). More recently Wherry's *The Fern Guide* (1961) revised and enlarged from the earlier *Guide to eastern Ferns* (1937) to include the mid-western states, and *The Southern Fern Guide* (1964), deserve special mention.

A rather remarkable number of state fern floras and annotated lists have been published, especially since 1900, so that most of the states now have some special publication on ferns. Blake (1941) lists over 150 fern floras and florulas. Several of these are books with detailed accounts of the distribution and ecology of the species, as well as keys, descriptions and illustrations, and they have greatly increased our knowledge.

Technical studies on American ferns, after the work of Eaton and Engelmann, received impetus from new explorations in the United States and tropical America. Elizabeth G. Britton, wife of the director at the New York Botanical Garden, published on the Bolivian collections of Rusby in 1888 and of Bang in 1895, and the second monograph, "A revision of the North American species of *Ophioglossum*" in 1897.

Professor Underwood suggested the idea of a "North American Flora" to Britton, and in 1896 joined the staff of the New York

Botanical Garden and initiated work on the Pteridophyta for that project. Underwood published some eleven revisions on fern genera, including *Phanerophlebia, Danaea, Stenochlaena* (= *Lomariopsis*), and the ternate *Botrychiums*. Most of his monographic studies were published in his series, "American ferns, I-VII" (1898-1907). He contributed the Pteridophyta treatment in Britton and Brown's *Illustrated Flora of the United States and Canada and the British possessions* in 1896, and two years after his untimely death in 1907 at the age of 54 his contributions were published in vol. 16 of *North American Flora,* the Ophioglossaceae jointly with Benedict, and the Marattiaceae completed by Benedict. Underwood attempted to define natural species and natural genera, and with his field experience, geographic perspective and good collections he was often successful in these efforts. He was outspoken about some of the earlier work which he considered to be overly conservative; for example, in reference to the *Synopsis Filicum* of Hooker and Baker he wrote (1902) : "Kunze's species have been slaughtered wholesale and many of them will have to be revived, and the same is true of many of the species of Fée." His points were generally well taken although he also fell into similar "errors" when his materials failed to be adequate to his demands on them.

Copeland went to the Philippines in 1903 to the newly activated Bureau of Science at Manila under Merrill. He began there a life-long interest in paleotropical ferns, later continued at Berkeley, that established his authority in that region. He published numerous floristic treatments, papers on novelties, and over twenty monographic studies. His major floristic work was the "Fern flora of the Philippines" (1958-60) in which he treated 157 genera and 934 species. His well known *Genera Filicum* (1947) was an amplification and maturation of his earlier "The oriental genera of Polypodiaceae" (1929) and was the first detailed account of all fern genera since *Natürlichen Pflanzenfamilien* (1898-1900). Copeland found in the Malaysian-Pacific region an exceedingly rich fern flora, poorly known when he began his work. His publications were effective in organizing much of this flora and they will remain basic to further studies. An evaluation of his *Genera Filicum* must await a time when the classification of the Filices is in less disarray.

Maxon graduated from Syracuse University in 1898 and, after a year of graduate study with Underwood at New York, joined the staff of the Smithsonian Institution. He collaborated with Underwood on the single number of the *North American Flora* devoted to ferns, furnishing the text for the Schizaeaceae, Gleicheniceae and Cyatheaceae (*Cyathea*). He started his own classical series, "Studies of tropical American ferns" (1908-22). These papers contain most of Maxon's 25 monographic studies. Those on sections of *Polypodium, Asplenium,* and the Cyatheaceae are especially notable. Maxon also supplied

excellent accounts of the Pteridophyta for manuals such as the second edition of Britton and Brown's *Illustrated Flora* (1913) and Abrams' *Illustrated Flora of the Pacific States* (1923), and for floras such as Tidestrom's *Flora of Utah and Nevada* (1925), and Kearney and Peebles' *Flowering Plants and Ferns of Arizona* (1942). His treatment of the Pteridophyta of Puerto Rico and the Virgin Islands (1926) is his outstanding work on tropical America. It was so carefully prepared that its usefulness has not diminished in the forty years since its publication. Maxon established a firm and broad foundation for the taxonomy of neotropical ferns. His finely and carefully drawn definitions of species often revealed ecological and geographic differences which provided a better understanding of relationships.

Morton has continued the floristic and monographic traditions at the Smithsonian. He contributed the treatment of Pteridophyta to Kearney and Peebles' *Arizona Flora* (1951) and to Gleason's *New Britton and Brown Illustrated Flora of the Northeastern United States and adjacent Canada* (1952). His monographic studies in the difficult Hymenophyllaceae and the Thelypteroid ferns have been of special importance, for example, "The American species of *Hymenophyllum*, Section Sphaerocionium" (1947), "On the genus *Cyclodium*" (1939), and "Some West Indian species of *Thelypteris*" (1963).

The tradition in pteridology at Harvard University, initiated by Gray and Eaton, was taken up by Fernald and Weatherby about 1915. Fernald critically studied many of the northeastern ferns during preparation of the Pteridophyta for his eighth edition of *Gray's Manual* (1950). He had a keen appreciation of geographic variation, especially when accompanied by isolation. This led him to recognize vicarious species such as *Asplenium cryptolepis* and *Pteretis* (= *Matteuccia*) *pensylvanica* and geographic varieties such as *Phyllitis scolopendrium* var. *americana* and *Polystichum braunii* var. *purshii*. Through his geographic studies Fernald brought a new viewpoint to the classification of ferns, and provided the basis for refinements in their taxonomy and the recognition of evolutionary complexes.

Weatherby was engaged in studies of the Connecticut flora when he first began to consult the collections at Gray Herbarium about 1908. In 1929, when he was appointed Assistant in the Gray Herbarium, he and Mrs. Weatherby took up residence in Cambridge. His influence extended far beyond his own publications in pteridology. He encouraged and guided many studies, for example, Maurice Broun's *Index to North American Ferns* which with his aid and professional advice was completed in 1938. Broun's *Index* recognized 333 native species of Pteridophyta from the United States. This is about the same number, but not quite the same species as are generally recognized now. Weatherby was concerned with the original materials of classical authors and published a definitive study of the type specimens of

Desvaux's ferns (1939) · He prepared, with the aid of Mrs. Weatherby, the first extensive series of photographs of fern types, most of them from the rich collections at Paris. He published several revisions of difficult groups in the genera *Polypodium, Asplenium, Selaginella* and *Notholaena,* and in most of these papers he effectively employed the geographic variety to bring out significant variations in species. See especially "The group of *Polypodium lanceolatum* in North America" (1922) and "The group of *Polypodium polypodioides*" (1939). His publications on tropical complexes did much to extend the geographic concept in fern taxonomy beyond the boreal and temperate regions and to dispel the preoccupation with the species level so widespread among tropical workers.

Monographic and floristic work has vigorously continued in the United States since 1950, with emphasis on tropical America. The development of stronger collaboration with botanists in Latin America has resulted in extensive new materials from the American tropics and has provided a better basis for monographic work and a greatly expanded basis for evolutionary and biogeographic studies. The publication in Britain of Professor Manton's *Problems of Cytology and Evolution in the Pteridophyta* (1950) supplied the impetus for American work in this field and has opened the way for critical investigations of variation and the process of speciation. Work on polyploid complexes, breeding systems, and details of the life-cycle are providing new knowledge on the biology and evolution of ferns. These experimental and field studies are bringing new information to the classification of ferns and are establishing a new tradition in American pteridology.

PALEOBOTANY

CHESTER A. ARNOLD

(*University of Michigan*)

Until 1865 and for a few years afterward, John Strong Newberry, Lesquereux, and John William Dawson were responsible for most of the paleobotanical research in North America. Newberry was a geologist whose interest in paleobotany developed during boyhood when he collected fossil plants from his father's coal mine in Ohio. Subsequently his interests extended to Mesozoic and Cenozoic floras from the western part of the United States. Lesquereux, an emigrant from Switzerland, also spread his efforts far along the geologic column, and he was also a bryologist of note. Dawson (known as Sir William) was a Canadian geologist noted mostly for his work with Devonian plants, but he included in his studies fossils from the United States. He also devoted some attention to Mesozoic and Cenozoic plants later in his career. His *Geological History of Plants* (1888) was the first paleobotanical textbook by a North American author.

When extensive coal mining developed in the eastern United States during the early 1800's, Carboniferous plants were unearthed in large numbers. Some of these were studied by Lesquereux whose efforts culminated in the publication by the State of Pennsylvania of the 3-volume *Coal Flora* (1879, 1880, and 1884). This was an outstanding achievement fully equal in significance to the contemporary works of Stur in Europe. Even today, after nearly 90 years, it remains the most comprehensive account of American Carboniferous plants.

After publication of the *Coal Flora* interest in American Carboniferous plants dropped rather precipitously, and never since has it gained the status achieved under Lesquereux. The withering of interest in the subject was probably due to the size of the *Coal Flora*. With its 111 plates it conveyed the impression that the task was finished and nothing of consequence remained to be done. Geologists gave their attention to the large and diversified invertebrate faunas that proved to be so useful in stratigraphic correlation, and the relatively few botanists of the day were completely engrossed in the study of living plants.

An attempt was made in 1886 to forestall the complete demise of American Carboniferous paleobotany by the appointment of David White to the newly organized United States Geological Survey. For a while it appeared that some of the activity reminiscent of past decades

might be resumed when White began to collect actively in the coal mining areas in the eastern half of the country. This phase of his activity resulted in the publication in 1899 of his *Fossil Flora of the Lower Coal Measures of Missouri*. This volume, with its 73 quarto size plates, ranks second to the *Coal Flora* as a contribution to American Carboniferous paleobotany. However, after a few years, White showed an aptitude for administrative work, and during the remainder of his career much of his time and energy was channeled into activities other than paleobotanical research. He did make some major contributions after 1900 but most of his papers were short. When he died in 1935 a vast store of knowledge of American Carboniferous plants went with him. Fortunately, he did leave several manuscripts that were sufficiently complete for posthumous publication. The longest of these, on the Lower Pennsylvanian flora of Illinois, is being prepared for publication by Charles B. Read.

In 1923 a new phase of Carboniferous paleobotany was set in motion in the United States when Noé reported the discovery of coal balls in Illinois. Up to then coal geologists, including White himself, emphatically maintained that coal balls did not exist in American coal-bearing formations. However, they were soon shown to be present not only in Illinois, but in Indiana, Iowa, Kansas, and elsewhere. Noé, therefore, earned deserved and lasting fame with his discovery.

During the decades between 1880 and 1920 when Williamson, D. H. Scott, Bertrand, Renault, and Solms-Laubach were adding volumes to our knowledge of the structure of Paleozoic plants, very little research on this subject was being carried out in the United States. The discovery of coal balls in Illinois changed this situation drastically. Within the last forty years research on plants in coal balls has reached major proportions and many significant contributions have been made.

The announcement by Kidston and Lang in 1917 of the plants in the Rhynie Chert started a worldwide upsurge of interest in Devonian paleobotany. This set the stage for a rather spectacular discovery in eastern New York State in 1920, which was the unearthing of about 200 *in situ* stump casts in a Middle Devonian sandstone bed near Gilboa in the Catskill Mountains. These stumps, along with the associated plant remains, were the basis for Winifred Goldring's reconstruction of *Eospermatopteris,* which at that time was interpreted as a pteridosperm. Though the Gilboa plants were quite different from those found earlier at Rhynie and were preserved in a totally different way, they were a striking foretaste of possible discoveries still to be made in the thick sequence of Devonian rocks in New York State.

When emigrants moved into the vast sparsely populated areas of the western part of the United States during the nineteenth century, they saw strewn over the ground pieces of rock bearing leaf imprints. Many large collections of these were subsequently made, some by

geologists making preliminary reconnaissance surveys of the country, and some by guides, explorers, trappers, and even by soldiers stationed at lonely outposts to protect travelers from hostile Indians. They also saw much silicified plant material including of course silicified wood. In the vicinity of the Black Hills they saw rough stony masses that to them bore more resemblances to beehives and wasp's nests than to anything else they could think of. These of course were the silicified cycadeoid trunks that had weathered out of the Lower Cretaceous Lakota formation and for which that region has become famous. At first scientists paid little attention to these fossils. No large collections were made until Lester F. Ward, Wieland, and others visited the region between 1890 and 1900.

In 1906 and 1916 Wieland described the Black Hills cyadeoids, along with several others, in two superbly illustrated volumes entitled *American Fossil Cycads.* The title, however, is a slight misnomer because it deals with cycadeoids, not true cycads.

The silicified tree trunks that the emigrants saw were in scattered localities extending from the Great Plains to the Pacific Coast. In age they ranged from Triassic to Pliocene. In a few places, as in the Yellowstone region, some of them still stood upright where they had grown millions of years before. We do not know who first saw these petrified trees but the oldest observer on record is the guide and scout Jim Bridger, who discovered Great Salt Lake about 1825. Bridger was illiterate and he left no written accounts but his fantastic stories became western folklore. For instance, he told of seeing petrified birds in the Yellowstone country perched on limbs of petrified trees singing petrified songs. Those he shot did not fall because gravity was petrified there! Mention of petrified trees also occurs in some of the writings of Washington Irving and Edgar Allan Poe.

The early explorers and emigrants marveled at the size of the stone tree trunks standing on some of the semi-arid hillsides in regions where trees at present thrive only around lakes and along streams. They correctly concluded that at some remote time during the past the country had received more rainfall than it did during the nineteenth century.

Some of the collections of Upper Cretaceous and Tertiary leaf impressions made during the days when the western United States was being explored and settled fell into the hands of Newberry and Lesquereux. Newberry's first publication on these plants appeared in 1861, and his last was a posthumous work published in 1898. The most spectacular achievement was again by Lesquereux whose 3-volume *Contributions to the Fossil Flora of the Western Territories* appeared in parts in 1874, 1878, and 1873. This work contains figures and descriptions of hundreds of Upper Cretaceous and Tertiary plants. Lesquereux also completed a separate study of the flora of the Dakota

Sandstone, which at present is the only comprehensive account of this great flora. The monograph describing it was not published until three years after his death in 1889.

Great strides were made in Mesozoic and Cenozoic paleobotany during the three decades following 1880, and these were due in large part to the organizing efforts of Lester F. Ward who had joined the United States Geological Survey. Several massive documents on Mesozoic and Cenozoic plants appeared as monographs and annual reports. Some were written by Ward himself but others were by Knowlton, Fontaine, Hollick, and Newberry. Some of these contained hundreds of pages and were superbly illustrated. The bulk of what we know today of American Triassic, Jurassic, and Lower Cretaceous plants was assembled then. Out of these efforts also came what may be called the "Index Kewensis" of American paleobotany, which was Knowlton's *Catalogue of the Mesozoic and Cenozoic Plants of North America* which appeared as Bulletin 696 of the U. S. Geological Survey in 1919.

A notable achievement was also by Hollick and Jeffrey who collaborated in a unique study of the lignitized coniferous remains from the Lower Cretaceous Raritan Formation of Staten Island. The results were published in 1909. These authors developed ingenious and original techniques for embedding the material for microtome sectioning, but they are so laborious that few subsequent paleobotanists have had the courage and perseverance to emulate them.

Berry, the most prolific of American paleobotanists, began to publish in 1903. Much of what we know about Cretaceous and Early Tertiary floras of the eastern and southeastern United States came from his efforts.

The early work on western American Mesozoic and Cenozoic paleobotany was performed before the land surveys were complete and when transportation was slow and difficult. Collectors did not always keep accurate records of where their fossils were found, often simply designating them according to the general area, i.e., a certain river valley, or so many miles from a fort or settlement. Moreover, specimens from different geological horizons were often lumped together without identifying marks of any kind. The inevitable result, of course, was collections that were badly mixed and of uncertain source.

The confusion was added to by the human element and the nature of the fossils themselves. Many of the collections consisted mainly of leaf imprints of deciduous trees. Paleobotanists tried to assign them to living genera, but this was difficult because of the few large herbaria in the country at that time which the investigators could consult. The result was that most of the fossils were placed in genera with which the investigators happened to be familiar. Most of these of course were represented by the deciduous trees that grew along the streets of New

York and Washington where much of the work was done. Some of the favorite genera of the paleobotanists of that era, regardless of the age or source of the fossils, were *Acer, Ficus, Laurus, Magnolia, Platanus, Populus, Quercus, Salix, Sassafras,* and *Vitis.* Knowlton's catalogue, for instance, lists about 150 fossil species of *Ficus,* which is certainly far in excess of the actual number represented by the fossils.

As a result of the chaotic state into which American paleobotany had fallen during its early years, much distrust developed among botanists and geologists toward the whole subject. They felt, and not without justification, that fossil plants did not "make sense." This attitude still persists in some quarters even today, but the situation has improved considerably during past years. Subsequent research has amply shown that the difficulties lay in faulty interpretations rather than in any inherent flaws in the fossil plants themselves.

The improvement referred to in the paragraph above began shortly before 1920. As might be expected, progress was slow and seemingly insignificant at first, and the whole task of setting the paleobotanical house in order is far from complete today. The first important step in the right direction was taken by Berry in his analysis of the Early Eocene Wilcox flora in 1916. This flora, which is widely spread throughout the states bordering the Gulf of Mexico, was found to show significant differences throughout its range. The eastern extension showed obvious resemblances with the living flora of the Caribbean region, while further west it showed similarities with living floras of Mexico and Central America. This was only a beginning, but it showed that ecological interpretations might be applied to fossil floras.

An individual who played a major role in unraveling the mess into which American Cretaceous and Tertiary paleobotany had gotten itself was the late Roland W. Brown. One of his major concerns was the correct identification of leaf imprints. He visited old localities and looked for more and better specimens. While doing this, he kept a sharp lookout for seeds and fruits that would often furnish valuable clues to the identities of some of the leaf impressions. He also made wide use of herbarium facilities available to him. Brown also emphasized the importance of making collections that were as large as possible in order that the natural variations within species might be manifest. He set many good procedural patterns that other paleobotanists could follow with profit.

R. W. Chaney, probably the leading contemporary investigator of western American Cenozoic floras, and several of his students concerned themselves with improved methods of identification, but they also made great strides in understanding the ecological relationships of fossil plants. They found, for example, that from leaf characters alone one could often form definite conclusions concerning the type

of climate under which plants grew during the past, and some knowledge of the probable climate often aids in identification.

Much of the confusion that formerly existed concerning the geologic history of the Taxodiaceae was cleared with the discovery of *Metasequoia* during the early 1940's. We now know that many of the formerly supposed fossil remains of *Sequoia* and *Taxodium* are properly *Metasequoia* and that this genus thrived in North America until Miocene time.

Better records of the sources of material, more satisfactory means of identification, and taking into consideration the probable climatic relations of fossil plants has made possible analyses of many North American Cretaceous and Tertiary floras that defied understanding 40 years ago. We know today that Tertiary floras were not uniform through time and space, and that they underwent massive migrations and changes in response to major past climatic fluctuations. While these achievements may be contemporary rather than historical, most having come about during the last 40 years, reference to them seems justified because they have supplied us with a massive body of information on the Cretaceous and Tertiary vegetational history of the North American Continent.

It is gratifying that some of the largest and most spectacular of the western American petrified tree trunks have been momentarily saved from the destruction threatened by vandals and commercially motivated individuals. Deserving of special mention are the logs of Late Triassic age in the Petrified Forest National Park in Arizona, the Eocene age "forests" in Yellowstone National Park, and the Ginkgo Petrified Forest, of Late Tertiary age in Washington State. There are also several smaller reserves, some of which are privately owned.

The science of palynology is of too recent origin to fill an important niche in a historical account. Yet the fact seems worthy of mention that the name *palynology* was originally proposed in 1944 in *Pollen Analysis Circular* No. 8, a mimeographed bulletin edited at Oberlin College by Sears. Also of some historical significance is the fact that the first paper inthe modern literature on Carboniferous spores was published in 1929 by the late Professor Harley H. Bartlett of the University of Michigan. In this paper Bartlett described three megaspore types that he assigned to Reinsch's genus *Triletes,* and in so doing demonstrated the feasibility of binomial nomenclature for fossil spores. The spores were found in coal pebbles buried in the glacial drift at Ann Arbor. Since 1929 the study of isolated spores and pollen has developed at a phenomenal rate and the investigations have embraced the geologic column from Cambrian to Recent. The volume of resulting literature is now large.

The interdisciplinary nature of modern systematics makes the subject a challenge for professors. I know very few, if any, who would dare claim to have successfully woven together the various threads of comparative botanical knowledge into a truly well-balanced synopsis of present-day thinking. The ideal systematist would have to be a "chemocytohistomorphotaxonometrician" — a rare bird!

WARREN H. WAGNER JR.

TAXONOMY

Charles B. Heiser, Jr.

(Indiana University)

After the death of Asa Gray in 1888, no one man was ever again to dominate the American botanical scene as he had. The period following his death until well into the next century was to witness little change in plant taxonomy. Botanical exploration and collecting, floristic and some monographic work were to occupy the energies of most systematists. Sereno Watson, and later B. L. Robinson and Fernald carried on the Gray tradition at Harvard. E. L. Greene was to continue to be the principal figure in the West for some years, followed by Jepson and Abrams. During the last years of the century N. L. Britton founded the New York Botanical Garden, the main emphasis of which was to be the production of floras, not only for the United States but for tropical America as well. The Field Museum of Natural History was founded in 1893, so at the close of the century all the major herbaria in the United States—the University of California, Field Museum, Harvard, Missouri Botanical Garden, New York, the Philadelphia Academy of Natural Sciences, and the Smithsonian Institution—were in existence.

Two other important events of the final few years of the last century must be noted here. In 1897 Charles Bessey sketched the outlines of his "phylogenetic" system which in a sense marked a maturing of American taxonomy. Prior to this date no American botanist had proposed a system of classification. There were, of course, ample reasons for this state of affairs—the great mass of undescribed species and the vast areas in the need of floras required all the attention of taxonomists. "Bessey's system," to be more fully developed in 1915, was never to compete with the Englerian system for the arrangement of plants in floras or herbaria, but it did have a strong following of teachers of systematic botany in the United States. Some of his "dicta" of "generally accepted practices of classification," with some modifications, are still of significance.

The last few years of the century also saw the first publications of Liberty Hyde Bailey, who was to become an outstanding leader in American botany. Here, emphasis should be placed upon his contributions to the taxonomy of horticultural plants, particularly his *Cyclopedia of American Horticulture* (1907-09) and his *Manual of Culti-*

vated Plants (1923), which in revised editions, are still the principal works in the field.

The rediscovery of Mendel's work and the flowering of the other branches of biology at the turn of the century were to have little immediate effect on plant taxonomy. Although Clements stated a need for an "experimental taxonomy" in 1905, the "new taxonomy"[1] did not actually come into being until the twenties. Harvey Monroe Hall of the University of California and later of the Carnegie Institution of Washington, who, upon occasion, collaborated with Clements, was to play one of the leading roles. Their contribution entitled "The Phylogenetic Method in Taxonomy" appeared in 1923, and although their method was eventually to have considerable impact on plant taxonomy, the use of the word "phylogenetic" in this connection can still engender rather heated discussions among present-day taxonomists, particularly between Americans and Europeans. Although in places this work was marred by Clements' neo-Lamarckian views and an ultra-conservative species concept, much of the work had a distinctive modern flavor. For example, in the introduction they state, "To be both comprehensive and thorough, taxonomy must draw its materials from all other fields, just as it must serve them in turn. While it leans most heavily upon morphology, it can not afford to neglect histology and physiology, and it must learn to go hand in hand with ecology and genetics in the future. Indeed, if it is to reflect evolution as accurately as it should, it must regard physiological adjustment as the basic process, and morphological and histological adaptations as the measurable results. This means that the taxonomist of the future will think in terms of evolutionary processes, and will learn to treat his morphological criteria as dynamic rather than static."

A year later appeared "*Hemizonia congesta,* a genetic, ecological and taxonomic survey of the hay-field tarweeds," by Babcock and Hall. Although not all their conclusions have been borne out by subsequent studies, this paper is one of the first truly "biosystematic" contributions with its emphasis upon chromosome numbers and artificial hybridizations.

At the same time important advances in systematic botany were being made in Europe. Turesson's now classic work dealing with transplants and ecotypes appeared in 1923, and, indeed, this work is

---

[1] The words, "new systematics," "experimental taxonomy" and "biosystematics," have been criticized by some systematists. There was in a sense no "new taxonomy" but simply an expansion of the "old" with the addition of the application of new methods and new ideas from the other developing fields of biology. Although I shall use the term biosystematics here, it is of interest that its use seems to be becoming less frequent in the literature, perhaps because of increased use and understanding of "biosystematic methods" which can now be thought of as belonging to the general field of systematics.

cited by Hall and Clements. Another European was to have an even greater direct effect upon American botany, for Jens Clausen, who in the twenties published on chromosome numbers in *Viola,* in 1931 joined the Carnegie Institution of Washington where, along with Keck and Hiesey, the work on transplants was greatly expanded and the ecotype concept was further developed. Over all, the work of this team had considerable influence on the newly developing field of biosystematics.

The events thus far related in regard to the "new taxonomy" were largely confined to California, but elsewhere in the United States important developments were occurring. In 1928 Edgar Anderson of the Missouri Botanical Garden published one of the first extensive analyses of two closely related species in an attempt to answer the question, "What are species and how do they originate?" Although some of his conclusions may not be acceptable today, this paper is particularly significant for the use of population samples, or "mass collections" as he was later to call them, and biometrical methods. The ideographs which he utilized here were only the first of a number of pictorial biometrical devices he was to introduce which would be extensively used by taxonomists. Although Anderson was to participate in it only to a limited extent, there was shortly to be an intense pre-occupation of geneticists, and to a lesser extent of taxonomists, with definitions of species which was to subside only in recent years. Anderson, however, was to have considerable influence in other ways.

In the late twenties "experimental" taxonomic work in the British Isles was commencing with the work of Marsden-Jones, Turrill, and Gregor, and there was to be an increasing interchange of ideas and methods between European and American workers from this time on. Of equal importance was the beginning of a greater interaction of zoological and botanical systematists in the United States to their mutual benefit. Kinsey, with whom Anderson had become associated in his years at Harvard, had begun his studies of gall wasps and there was a reciprocal influence between the two. Other zoologists, Dobzhansky prominent among them, were to have a significant influence on plant taxonomists in the next decade.

The taxonomic application of an understanding of polyploidy began in the thirties. Winge of Denmark in 1917 had formulated a hypothesis of the origin of species through chromosome doubling, and the hypothesis was verified experimentally in 1925 by R. E. Clausen and Goodspeed. In 1931 the Canadian Huskins postulated that *Spartina townsendii* arose from hybridization of *S. alternifolia* and *S. stricta* followed by chromosome doubling. Then in 1936 Anderson produced strong circumstantial evidence that *Iris versicolor* arose from *I. virginica* and *I. setosa* through such a process and showed that studies of polyploids could be an aid in interpreting earlier distribution of species.

The understanding and appreciation of the role of hybridization and its frequency in higher plants were to begin with Anderson and Hubricht's study of *Tradescantia* in 1938, to be followed by Anderson's introduction of the concept of introgression. This work was followed by a great number of studies of hybridization in plants, and, more recently, in animals as well.

Although the phenomenon of apomixis had been recognized and understood for some years, it was not until the year 1938 through the work of Babcock and Stebbins of the University of California, and their recognition of "agamic complexes" in the western North American species of *Crepis,* that much headway was made with the systematic treatment of apomicts.

Thus with the development of population sampling and an understanding of the taxonomic significance of polyploidy, hybridization and apomixis, systematists were well on their way to solving what up to that time had been some of the most difficult taxonomic problems at the species level. In the forties biosystematic studies were pursued by an increasing number of systematists. Stebbins, *Variation and Evolution in Plants* (1950), while not strictly speaking a taxonomic work, deserves mention for it was a masterful synthesis and has had profound influence on the newer generation of plant taxonomists.

While the challenge and appeal of the developing field of biosystematics attracted a large number of taxonomists from the late 1930's on, floristic and standard monographic work, of course, did not cease. It is of interest to note, however, that what is generally recognized as one of the outstanding floras of the United States at the time, Deam's *Flora of Indiana* (1940), was produced by a man who had not been trained professionally as a botanist. The fact that today many areas of the United States have no manuals, or at least no up-to-date manuals, for the identification of their floras has been of grave concern, if not embarrassment, to many American taxonomists. Recent years, fortunately, have seen considerable renewed interest in floristics, and several excellent state and regional floras have appeared.

There has never been a floristic work that treats the whole of the United States. Such a treatment had been the aim of both Asa Gray and Britton but their efforts never saw completion. In 1966 a committee of American taxonomists set up the "Flora North America Project" with the aim of producing a concise, diagnostic treatise of the vascular plants of the continental United States, Canada, and Greenland. This work, obviously a long term endeavor, will involve the cooperation of many taxonomists and should be of immense usefulness.

In addition to the work in biosystematics and floristics there have been several other noteworthy developments in systematic studies in the United States. The use of anatomy and related disciplines as an aid in systematic studies has increased greatly in the last few decades. The studies of I. W. Bailey of Harvard and several collaborators beginning

in 1942 on the woody ranalian genera rank among the most significant in the field, and served to focus attention on the importance of studies of this nature.

For some time after the publication of Bessey's system in 1915 there was little concern on the part of American taxonomists with the phylogeny and general systems of classification of the angiosperms. Recently there have been renewed activities in this area with the systematists able to make use of a large body of new data bearing on the subject. One of several taxonomists who have been active in this area is Arthur Cronquist of the New York Botanical Garden, whose *The Evolution and Classification of Flowering Plants* (1968) is indicative of the current interest.

As a tool for solving many taxonomic problems, chemical taxonomy has already demonstrated its importance. With the death of Ralph Alston of the University of Texas in 1967 at the age of 41, the chemical taxonomic movement lost one of its foremost spokesmen and contributors. However, the field is now well established and currently one of the most active in plant systematics in the United States.

Another recent development, numerical taxonomy, on the other hand, in spite of several enthusiastic pioneers, has not yet had any great impact on the studies of the angiosperms. Although there is growing awareness of the importance of the use of computers as an aid to systematic work, there has been a reluctance on the part of the plant taxonomists to accept the dogma of numerical taxonomy as advocated by certain zoologists.

It is obviously impossible to mention all of the significant contributions. This is one person's view of some of the more important trends in plant taxonomy, and attention should be called to the comprehensive reviews of Constance (1955, 1964) and Keck (1957).

PLANT GEOGRAPHY

Joseph Ewan

(*Tulane University*)

Plant geography is to vegetation what etymology is to language: it lays bare the patterns that provide the symmetry. And just as etymologists may disagree on word origins, so plant geographers frequently find different explanations for the patterns they encounter in the world's floras.

The first American item of phytogeographic interest was an addendum Humboldt made to his *Essai sur la Geographie des Plantes* ("1805," 1807) in which Professor B. S. Barton is reported as having noticed anomalous plant distribution east and west of the Alleghany Mountains. Barton was visited by Humboldt and Bonpland in Philadelphia on their return from tropical America to France, and the memoir to which Humboldt referred was finally published in 1809, *A Specimen of a geographical view of the Trees and Shrubs, and many of the herbaceous Plants of North America, between the Latitudes of seventy-one and twenty-five.* Humboldt's classic diagram of the zonation of plants was adapted for temperate latitudes and redrawn to serve as a frontispiece for the first edition of Mrs. Lincoln's *Familiar Lectures on Botany* (1829). The caption reads "Distribution of plants in equinoctial America according to elevation above the level of the sea." Mrs. Lincoln's mentor, Amos Eaton, does not comment on Humboldt in his introductory botany texts.

Dr. Pickering, who had been tramping the White Mountains of New England with Oakes, published a short article in 1827 "on the Geographical Distribution of Plants." Rafinesque commented that he did not admit of all of Pickering's conclusions, nor did he think his map quite correct, "yet he has opened the way." It may have been Pickering who contributed the anonymous account of plant geography to the Philadelphia edition of Murray's *Encyclopedia of Geography* (1837, 1:236-254, and 2:406-452). The author was masked in the preface as a "gentleman of high reputation in the scientific world" who adapted Hooker's text prepared for English readers with American examples. When Gray wrote an obituary of Pickering he alluded to "a brief essay on the geographic distribution and leading characteristics of the United States flora, which few of our day have seen." The horticultural writer Lippincott (1864) referred to this anonymous

article without identifying its author though he could have known Pickering in Philadelphia.

Rafinesque summarized "botanical geography and localities" in ten pages in his *New Flora of North America* (1836) recognizing seven floristic regions where deCandolle distinguished three:

1. Boreal Region
2. Canadian Region
3. Alleghanian Region
4. Floridian Region
5. Louisianian or Missourian Region
6. Texian Region
7. Origon Region

Among twelve questions he posed two still persist: "Why do the tropical genera so seldom extend into Florida?" and "Why are the two shores of North America, east and west, so unlike to each other in vegetation?" Rafinesque leaves the topic with this tease: "These queries and others of a similar nature may exercise the ingenuity of speculative Botanists, or amuse their idle hours; but they are facts and as such deserve our notice." In the same year Meyen's *Pflanzengeographie* (1836; Ray Soc., 1846) carried references to the United States, but the sources are too out-dated to be useful.

Surgeon Hinds, Royal Navy, declared that "Her Majesty's Ship *Sulphur* was the school in which I more particularly studied geographic botany." In the tradition of Darwin aboard the *Beagle,* Hooker on the *Erebus,* and Huxley, the *Rattlesnake,* Hinds turned his voyage of circumnavigation to biological advantage. He appended to volume two of the *Narrative of the Voyage* (1843) "Regions of vegetation being an analysis of the distribution of the vegetation forms over the surface of the globe in connexion with climate and physical agents." He acknowledged using the Philadelphia edition of Murray's *Encyclopedia of Geography,* containing what may have been Pickering's article, and recognized ten regions for North America in his short penetrating account. Hind's "California Region," however, extended east to the Rocky Mountains, and he postulated that "however interesting the Rocky Mountains may prove to the geologist, they have no flora sufficient to give them any individuality as a region."

Asa Gray did not consider plant geography in his *Elements of Botany* of 1836 but his competitor Alphonso Wood devoted a few pages to the subject in the first edition of his *Class Book of Botany* (1845) stressing causative factors. In 1840 Gray turned botanists' attention to a comparison of the floras of Japan and North America in the course of reviewing Siebold's *Flora Japonica.* In that memorable year, 1859, Gray published a detailed analysis of the floras of Japan and North America based on Wright's collections. He listed 580 entries for Japan with their floristic affinities, and found 61% showed a Sino-American

relationship. The flora of western China, with its even closer affinities with North America, was uncollected at that time, hence unknown to Gray. Gray continued to harrow the topic, and his presidential address before the AAAS, published in its *Proceedings* (1873) is often acknowledged a classic. In 1858 he had remarked before the American Academy of Arts and Sciences: "I cannot resist the conclusion that the extant vegetable kingdom has a long and eventful history, and that the explanation of apparent anomalies in geographical distribution of species may be found in the various and prolonged climatic or other physical vicissitudes to which they have been subject in earlier times." Hui-Lin Li (1952) of the Morris Arboretum succinctly reviewed Sino-American floristic relationships with the aid of 56 maps in an excellent epitome.

As a member of the Wilkes Exploring Expedition around the world Dr. Pickering gathered information for a book on which he worked for seven years titled "Geographical Distribution of Animals and Plants"—the last of the unhappy program of publishing the Expedition's findings: it was never published. Two sections were privately printed by Pickering: a short introduction to the whole (1854) and what amounted to a transcript of notebook records on plants and animals in their wild state taken down as they were observed or collected (1876). Both of these are scarce today.

"Observations on the necessity of a more enlightened agriculture" in the form of *Agricultural Reports of the United States Commissioner of Patents* were issued over a span of years on practical topics. The horticultural writer Lippincott published a discursive "Geography of Plants" as part of the Report for 1863 (1864) noting effect of climate on the regional growing of wheat and wine grapes. Historical notes from many sources are included.

Merriam's classification of Life Zones was born of the Great West. At seventeen, through Baird's influence, he spent the summer with the Hayden Survey in the Yellowstone region—John Coulter was one of the field party. Merriam's thesis that temperatures at the time of flowering and of breeding seasons of animals are decisive in the distribution of the organism originated evidently with zoologist Verrill during Merriam's student days at Yale where he was also a member of D. C. Eaton's class. The Bureau of Biological Survey, founded in 1885, and of which Merriam was made head, influenced American botany because of the genuine interest "Merriam's boys" took in botany as well as zoology. Merriam, a mammalogist, encouraged this: *Arctomecon merriamii* of Coville and *Arctostaphylos merriami* of Eastwood testify to the recognition of his interest. Issues of the *North American Fauna* series, which he founded, included a botanical introduction as a regular feature.

Merriam's first report (1890) gave the geographic distribution of

plants and animals of San Francisco Mountain in Arizona, the summit of which (12,655 ft.) is now known to support a disjunct alpine flora of Rocky Mountain affinities. Merriam reported his Life Zone observations with the object of giving as complete an exposition of conditions as possible rather than with the idea of proposing a theory. Interest in the West was high. Arizona Territory held a particular fascination, and Merriam's Life Zone report, easily comprehended even though the divisions of the Boreal Zone could not always be distinguished, made him in the words of his biographer Osgood, a "power in the land with a reach into posterity that will long be felt." For 61 years he wrote scientific articles.

Coville employed Life Zones in his classic *Botany of the Death Valley Expedition* (1893) where he introduced the term "zonal plant" for what later was called an "indicator species." Coville selected *Larrea,* for example, as diagnostic of the Lower Sonoran zone. In this report he first analyzed the phytogeographic affinities of the floras of the Death Valley and the Sierra Nevada. Merriam's scheme was taken up by H. M. Hall in his "Botanical Survey of the San Jacinto Mountains" (1902). Hall incorporated Life Zone data on his herbarium labels, thereby spreading Merriam's concept among field botanists. Abrams also took up the concept in his "Phytogeographic and taxonomic study of Southern California trees and shrubs" (1910). So accepted had Merriam's scheme become by 1915 when Hall and Abrams wrote botanical essays for the Pan-Pacific Exposition tour-guide, *Nature and Science on the Pacific Coast,* that the classification was used without notice of its author. The zoologist Grinnell who wrote on the California faunas in the Merriam language also collected hundreds of plant specimens in the course of his field work on the Pacific Coast. Phytogeographic commentary was a part of his zoological writings. In 1921 the botanist Smiley re-examined the postulates of the Merriam system and concluded that when trees and shrubs were selected as "indicator plants" they served a purpose, but that herbs and annuals proved of little value in the field. Jepson prefaced his *Manual* (1925) with an "outline of geographic distribution of seed plants in California," wherein he tidily incorporated a vast amount of detail, using Merriam's scheme as a frame of reference. Dansereau in his *Biogeography* (1957) considers Life Zones as "Bioclimatological" rather than "Historical." Yet his and such *schema* as J. G. Cooper's "Forest Regions" of 1859 and the "Floristic provinces" of Good (1953) have unmistagable ingredients of Merriam's classification. For example South Florida has the same areal circumscription in all three *schema,* that is, Tropical (Merriam), Floridian (Cooper), Caribbean (Good). Dice (1943) recognized South Florida as a well-marked subdivision within his Austro-riparian "Biotic province," on the Gulf Coast, but does not admit it as a separate province.

Harshberger's *Phytogeographic Survey of North America* (1911),

volume thirteen of Engler's monumental *Vegetation der Erde*, has 800 pages, is well indexed, but is unreliable and inaccurate, and so as an orientation for a topic must be verified against other works. Of all the branches of botany, Fernald wrote, plant geography demands "thorough training in exact taxonomic detail accompanied by the most discriminating judgment and prolonged and painstaking field study." His review of Harshberger's book in *Rhodora* (1911) ended with the warning that "any conclusions which may be innocently based by the unwary upon this 'Survey' will always be open to doubt."

Fernald by this time had gained an intimate knowledge of the flora of northeastern North America and had probed for origins and forces in its history. Next to the "Age and Area" theory of Willis, Fernald's classic "Persistence of plants in unglaciated areas of Boreal America" (1925), called the "Nunatak Theory" for short, aroused more discussion than any other phytogeographical thesis of this century. Merrill (1951) remarked that Fernald, "in reemphasizing the theory of persistence and in stimulating studies as to its implications may have made the largest single contribution to the science of phytogeography since the time of Darwin." That the botanists Butters and Abbe (1953) and Rousseau (1953), and the zoogeographer Wynne-Edwards offered contrary evidence to Fernald's thesis, striking sparks on granite, only emphasizes the character of the rock. Cain (1944) provided a well chosen instance of the effects of nunataks in relation to relics and senescence in *Iris setosa*, from the work of Anderson.

For over thirty years Professor Campbell of Stanford mapped his Sabbatical to the "steaming stillness of the orchid-scented glade" and to the "pile-built village where the sago dwellers trade" with an eye for hepatics in particular. From his notes and photographs grew *An Outline of Plant Geography* (1926), a well illustrated but undocumented cosy description. Campbell took a special interest in the biogeography of the Pacific and his "Origin of the Hawaiian Flora" (1918) and "Derivation of the Flora of Hawaii" (1919) were early landfalls on the subject. Later Fosberg sharpened the focus in a short summary included in the first volume of E. C. Zimmerman's *Insects of Hawaii* (1948).

Campbell's contemporary, Professor Setchell of Berkeley, also took a global view though his research interest was the geography of the pelagic algae of the North Pacific. Setchell fancied "geobotany" and his lectures opened vistas on the work of Skottsberg on antarctic problems, on long-distance dispersal and the importance of ecesis as a distributional factor in the "Law of the Minimum." His colleagues around the world joined in his Festschrift *Essays in Geobotany* (1936). One of Setchell's papers of far-reaching validity entitled "Temperature and anthesis," published in the *American Journal of Botany* (1925) concerns "waves of anthesis."

Geographically limital species often are the most similar morphologi-

cally was Epling's observation in studying the genus *Salvia,* subgenus Calosphace (1939). Matthew's hypothesis, put forward originally for fossil vertebrates, suggests that the "most ancient members of a group are not to be found in the old center of evolution but rather at the periphery of their migratory area," and appears to be cogent when applied to plants. Ewan (1944) reviewed the topic with special reference to the perennial Tolguacha (*Datura meteloides*) of the Southwest and northern Mexico, and proposed the term Matthew's Hypothesis of Peripheral Populations. Stebbins (1940) found the same pattern in the Sino-Himalayan genera of the Cichorieae, *Dubyaea* and *Soroseris.*

Weeds intrude into plant geography at every turn. The California flora continues to excite the interest of the cytogeneticist, paleobotanist, and ecologist, all of whose findings contribute to the interpretation of the origins and development of its flora. Yet in a short *thirty years* the introduced, adventive, and naturalized species in California have increased about 157% (*Leaflets of Western Botany* 9: 123. 1960). The Chilean tarweed *Madia sativa* arrived in California at least as far back as 1771 from the record of adobe bricks. The impact of Mediterranean weeds as competitors with native annuals since the Franciscan Mission Period in California has many facets. Influenced by a question raised some years before by Claypole of Antioch College, Asa Gray asked in a short provocative essay, "Pertinacity and predominance of weeds" (1879), "Why are there so many European weeds in northeastern North America, and so few American weeds in Europe?" Hultén's "milestone in botanical cartography" entitled *Amphi-Atlantic Plants and their Phytogeographic Connections* (1958) reopened the question. Gray was doubtful that European weeds were "more plastic" or that they had greater variability than American species. In any event, the speciation that has taken place in *Chenopodium,* for example, since its arrival in the seaports of the Atlantic Colonies, will surely be encountered in other weed genera. More data on the extent of selfing, and of apomixis, and more precise information of the role of birds as dispersal agents for weeds is urgently needed. Few subjects in plant geography hold the unpredictable place under the sun that weeds do: opportunistic from the hour of germination, adapting to new life (and death by herbicide!), weeds are *the* successful species of their times.

Livingston and Shreve's "Distribution of Vegetation in the United States as related to climatic conditions" (1921) issued as *Carnegie Institution of Washington Publication* number 284, will remain the classic attempt to find the one environmental value which determines the distribution of vegetation types. A companion survey was "Natural vegetation of the United States" by Shantz and Zon (1924). Zon covered the forested areas; Shantz, who was always interested in life's ecotone, the grasslands and deserts.

Shantz, like Clements, Shreve, and MacDougal, was a dusty-boots botanist of the West. During his years in Colorado he studied soils and vegetation and during two extended treks across Africa as a representative of the Allied Peace Commission in 1918-19, and of the Educational Committee to East Africa in 1924, Shantz again noted edaphic relationships and took about five thousand photographs of vegetation. A third of a century later he retraced his routes and rephotographed the vegetation at the sites, though 79% of the original locations were impossible in 1956-57 from agricultural or urban changes that had taken place. In 1958 a selection of 43 matching photographs was published by Shantz and Turner. A comparable "photodocumentation of vegetational changes in the northern Great Plains" based on Shantz's original records and using 52 paired photographs, was published by W. S. Phillips ,1963) after Shantz's death. Two other rephotographing projects, the Great Basin of Utah and the Arizona-Sonoran desert region, remain unpublished from Shantz's records, and are preserved at the University of Arizona. These time-lapse studies provide data for dynamic plant geography.

That polyploid species do not always have a wider range than diploids, and that polyploids do not cope more efficiently with inclement northern environments than diploid species, is shown by the 109 taxa of 22 genera of *Euphorbiaceae* surveyed by Perry (1943), and by the *Leguminosae* investigated by Senn (1938).

The Ozark-Ouachita Highlands exhibit disjunct colonies of Alleghenian species (*Arabis viridis, Acer saccharum, Phryma leptostachya,* etc.), and Mexican species or species with Mexican affinity, range northward from the Cordillera Oriental into southwestern Missouri (*Juniperus mexicana, Delphinium treleasei,* etc.). Hopkins (1937) suggested that the former represent relicts whose presence is controlled by a water requirement and that the latter are relicts along the margins of a Cretaceous sea whose presence is restricted by calcareous outcrops.

The endemic Venus Fly-trap (*Dionaea muscipula*) known for two hundred years, has long been living within the ecotone between the pocosin and savannah of southeastern North Carolina and adjacent South Carolina. But it was not investigated ecologically until 1954 when P. R. Roberts and Oosting began field and laboratory studies (*Ecological Monographs,* 1958). They found the chief factors delimiting the colonies to be the depth of the ground water table, characteristics of the soil surface, nutrient level, light intensity, and fire.

Stanley Cain's *Foundations of Plant Geography* (1944) is one lasting stake in the whole transect of literature. The book is divided into five parts: previous expositions; paleoecology; areography—here introduced as a term; speciation; and geobotany-polyploidy. It is documented with over 700 references and selected maps, and provided with a glossary. Cain left floristic description to others with the result we still have no *precis* for North America. The restriction that soils exert

over endemics is barely accounted for, as noted by Mason (1945) in his review, and the position which Cain takes that endemics must be either youthful or senile species is insecure since an obligate species may be stalled on a stable substrate. Mason and Cain believe that the "vagaries of mass interpretation of area are too great for their safe application to the interaction of these rather complicated phenomena with the events of history." Why such an excellent primer of plant geography has not been kept in print is a late Pleistocene riddle.

Perhaps the most important symposium in plant geography of recent years was held in Boston and reported in *Ecological Monographs* in 1947. The participants in the symposium on "the origin and development of natural floristic areas with special reference to North America" included leading plant geographers of the decade: Braun, Cain, Camp, Chaney, Just, Mason, and Raup. Camp presented several original maps of distribution based chiefly on the New York Botanical Garden Herbarium, and again directed our attention to the importance of the antarctic corridor in plant migration. Just discussed "geology and plant distribution" with the advantage of his familiarity with foreign literature. He admitted the bipolar evolution of conifers to be "quite probable," that "with few exceptions the old doctrine of historical plant geography is apparently still valid," and that paleobotanical data informs us on the history of modern plant families, but, in the words of Berry, "not of the founders of dynasties." Professor E. Lucy Braun's discussion of the "Development of the deciduous forests of Eastern North America" is a digest of her detailed book on that subject based on 25 years of field study. Cain confides that "natural areas" are "incapable of exact definition," whose very indefiniteness is one virtue. He concluded that Clement's monoclimax hypothesis is "not so much an ecological touchstone as a millstone," that the "single-factor operation does not occur in biological nature," and that "diverse analytic data cannot at present be synthesized back again into anything like the natural whole of the ecosystem."

Bipolar distributions of vascular plants between North and South America were considered in some detail by Bray (1898 and 1900) and by I. M. Johnston (1940) but the most comprehensive survey of "amphitropical relationships" was made by Raven in the *Quarterly Review of Biology* in 1963. Raven concluded that apparently the only explanation for these disjunct species or species-pairs, mostly herbs that have migrated from the north, is relatively recent long-distance dispersal.

International Studies on the Quaternary and *Quaternary of the United States*, edited by H. E. Wright, Jr., a geologist, and Frey, a zoologist, are outcrops of the Seventh Congress of the International Association for Quaternary Research, held at Boulder, Colorado, in 1965. Vegetational history is surveyed from a County Tipperary bog,

postglacial pollens in Finland, and Danish tills, to Carolina Bay sediments.

Without doubt the most dramatic phytogeographic event of this century was the discovery in China in 1944 of living trees of conifers earlier described as fossils. Shigeru Miki described the fossil *Metasequoia*, "Dawn Redwood," in 1941. Four Chinese botanists, Miki, Wang, Cheng, and Hu, rediscovered and recognized the fossil and modern occurrences of *Metasequoia*. That the North American fossils of Cretaceous and Tertiary age, long identified as species of *Sequoia* or *Taxodium*, were indeed this same Chinese conifer came with comparisons made by Chaney and summarized in the *Transactions of the American Philosophical Society* (1950). He studied the living *Metasequoias* and their associates growing in Szechuan and Hupeh in 1948. The talisman of the march of events was Merrill by whose distribution of seeds *Metasequoia*, "for the first time in nearly a score of million years, lived again in the western hemisphere."

PLANT ECOLOGY

PAUL B. SEARS

(Yale University)

Infants are usually named after they are born. So it has been with ecology. I would be inclined to place its formal beginnings in the United States a few years before Haeckel coined the word "ecology" in 1866. Asa Gray's *Flora of Japan* (1859) was called his *magnum opus* by Hooker. In this and his later work on forest geography and archeology Gray placed plant life in the perspective of time and process, taking note of environmental influences. His contemporary Marsh, though not a botanist, clearly understood the role of vegetation in maintaining health of landscape. His *Man and Nature* (1864), demonstrating that man had become a major geological influence, is a landmark in applied ecology.

This same decade saw the beginning of land grant colleges devoted to scientific agriculture, lending new challenge and standing to biology. Ruthless exploitation of forests and wildlife during the 1870's and 1880's brought out protests from scientists, with no visible effect. Forbes of the Illinois Natural History Survey anticipated the ecosystem concept in his 1887 paper on the lake as a microcosm, fostering the ecological viewpoint before the work of Schimper, Warming and Haberlandt introduced it to academic circles.

Meanwhile the system of liberal arts colleges was being modified by the continental university concept. Faculties and facilities were being assembled west of the Atlantic seaboard, often comparing favorably in everything except age and prestige with older institutions. The closing years of the 19th century saw a growing respect for science, greater freedom from traditional academic restraints and notable concern for plant life for esthetic as well as economic reasons. This concern was evident in improved agricultural research, forest policy and the national parks movement.

The work of the great European masters of taxonomy, structure, and function was readily accepted as it became known, since it represented an extension of orthodox botany, focussing on the individual plant and lending itself to laboratory methods. The same welcome was not extended to the ideas of the three European pioneers of plant ecology named above, save on the part of two youthful universities, Nebraska and Chicago. Here botany was under the leadership of men whose minds, to quote one of them, Bessey of Nebraska, had been kept in a meristematic condition.

124

Two of his protégés in this grassland state produced the classic *Phytogeography of Nebraska* (1898). One of them, Pound, later became a distinguished jurist and dean of the Harvard Law School; the other, Clements, tried to rationalize the enormously complex field of plant communities and wrote another classic, *Plant Succession* (1916).

Chicago is situated in the ecotone between the forest and grassland provinces and within the limits of the last glaciation. Here Coulter had assembled a brilliant group of botanists including Cowles who had studied earth science with Chamberlain and Salisbury. Stimulated by the work of Warming, Cowles showed in his study of dunes (1899) and physiographic ecology (1901) that plant communities are not merely the expression of physiographic process but active agents as well.

For a time these two universities were the chief sources of ecological training. To exemplify their influence and with no reflection on many other individuals of great ability we may list the following who have studied at one or both of these centers: Adams, Cooper, Fuller, Gleason, Hanson, Korstian, Livingston, Nichols, Pool, Shantz, Shelford, Transeau and Weaver. Students of their students are now making ecological history here and abroad.

At Nebraska the graduates of an ephemeral school of forestry were well grounded in plant ecology. At one time the majority of research stations of the U. S. Forest Service were headed by them.

As Gleason in his excellent paper "Twenty-Five Years of Ecology" (1935) has pointed out, the first phase of ecology was descriptive. Aided by the prevailing interest in taxonomy and stimulated by the extent and variety of the nation, it went beyond the simple floristic lists or inventories of the older naturalists. The rising group of ecologists were concerned not only to know what plants were present, but what activities and processes were taking place. It had become clear that plant communities were both products and agents of change.

As ecologic description progressed, passing into vegetation analysis or plant sociology, it developed problems of method, nomenclature, and theory, many of which continue to be debated. At issue is the degree of precision that is necessary or even possible in a given instance. More profoundly involved are the two approaches that must be used in facing any complex problem: dissection and analysis of components on the one hand, scrutiny of the whole in search of pattern on the other. Both of course are necessary and should be mutually corrective. But each has its advocates, according to temperament and environment. Cf. Sears' *Some Notes on the Ecology of Ecologists* (1956).

A definition, as the semanticist Ivor Richards has pointed out, is not a commandment, but rather an invitation to use a term in a certain sense. The same probationary value applies to mathematical formula-

tions of biological phenomena. American workers have been criticized for neglect of European approaches to plant sociology—a difference due in part to the longer apprenticeship and more intensive space conditions in Europe. The chief explanation is a pragmatic one rather than ignorance or wilful neglect; both techniques and terminology will be adapted as they clearly become necessary or even useful. Dansereau's orderly *Biogeography* (1957), demonstrating his impressive knowledge and wide travel, leaves no excuse for American neglect of alternatives, nor for lively differences as to the nature and classification of plant communities.

These differences have not so much changed since they were outlined by Gleason (*l.c.*) as they have been sharpened by increasing refinements of analysis. Certainly the idea of succession towards metastable climax as used by Cowles and Clements has provided a widely useful theoretical model for descriptive purposes. Whatever their actual origin or prospects for further change, most communities can be read as equivalent to some stage in an idealized successional series. This alone, at least in reconnaissance, represents a considerable economy of effort. But any model, too rigidly applied, can defeat its purpose.

Both as to the nature of plant community (association of Europe) and the question of climax, there are extremes of opinion. The community has been called a quasi-organism by some, a more or less random aggregate by others. As to the range of ideas concerning the climax, an extensive and thorough review has been published by Whittaker (1953). Not only are the relevant phenomena complex and difficult to measure, but their treatment involves problems that go back to the time of Plato, as to the nature of reality.

With improvements in instrumentation and techniques have come increasing efforts to quantify the study of vegetation. Striking advances in the physical sciences in recent decades have stimulated the use of models and systems analysis, reenforced by computer. Where these methods are employed with constant reference to the actual phenomena in question, the epistemic problem should come to be better understood than it has been.

The virtue of a model lies not in its finality, but in the fact that it is explicit and suitable for testing. Any tendency to apply the early concepts of vegetation and process too rigidly has been countered by the growing knowledge of soils and climate, as well as by experience with vegetation itself. The study of soil genesis that had begun in Russia was quickly expanded in the United States; Hilgard's work on the relation of soils to vegetation (1911) is an important ecological contribution, as are Thornthwaite's papers on the classification of climates. In broadening the idea of succession by emphasis on change, Cooper not only reacted against dogma, but helped to place the problem in

historical perspective. Similarly the view of vegetation as a continuum by J. T. Curtis has served to modify more static concepts of communities.

Notable among the many students of the great climatic formations are: Braun for the eastern deciduous forest; Clements, Weaver, and Shantz for the grassland province; and Shreve for the southwestern desert. Inevitably such studies have moved from static description to dynamic explanation in terms of physiological interaction with environmental forces, and on to genetic considerations in two senses of that word. For the individual plant its physiological amplitude is determined by its genetic endowment; and to account for any community as fully as possible, its antecedent history is needed. That is to say the study of vegetation has moved from a static, through a dynamic to a genetic approach.

Progress in the analysis of the dynamic phases of vegetation depends both upon refinements in plant physiology and upon understanding of the inanimate environment. Both have benefitted enormously from recent advances in instrumentation, but an excellent example of earlier work is that of Shantz and the physicist Briggs (1914-17) on the water requirements of plants. Later the laborious studies of Weaver on root systems were expanded from structure to function.

While the primary role of light in photosynthesis is of major ecological interest, so is the economics of energy as it is deployed, utilized and dissipated throughout the community. Although the American Gibbs had made distinguished contributions to thermodynamics, and its implications were understood by the historian Charles Francis Adams, Bayliss' *Principles of General Physiology* (1915) did much to create an interest in energetics among American biologists. Cannon's *Wisdom of the Body* (1932) made clear the importance of homeostasis within the organism; its role in the living community remained a challenging and difficult problem.

Transeau (1926) had discussed the energy budget, while a classic American study on the trophic-dynamic aspect of ecology was that of Lindeman (1942). Lindeman, unfortunately, did not survive long to continue his brilliant work which followed that of Juday by two years. Odum's review (1968) of developments in the field of energetics as applied to living systems discusses the contributions of contemporary workers that are beyond the scope of this paper. Incidentally the acceptance of Tansley's concept of the ecosystem has done much to clarify thinking.

Essential to an understanding of the concepts of succession and climax are the energy relations involved. While there is controversy as to whether any ecosystem exhibits metastability, and judgments involve a considerable degree of intuition, observation and experience encourage the view that living communities tend to approach an open

steady state, utilizing available energy not only to do work but to keep themselves in working condition. The obvious disruption of this constructive process is basic to the concern for conservation of natural resources, to be discussed below.

Ecology and genetics are, in a sense, twin disciplines. Both are implicit in Darwin's *Origin* and while not named are prophesied as inevitable in that work. Lending itself to precise experimental and mathematical controls and to economic application, genetics has advanced rapidly. For a time the remarkable plasticity of many plants under environmental change confused the understanding of variations, heritable and otherwise. Great credit is due to experimental taxonomy by contemporary workers in California and elsewhere, to studies of grassland ecotypes at Chicago, Nebraska, and Texas for applying modern genetics to species ecology, and to the review by Cain (1944) for reminding plant ecologists of the importance of this sister science to their field.

Corollary to the broadening concept of ecosystem dynamics and succession is an understanding of the historical factor. European pioneer reports on the vegetation changes revealed in peat were duly noted, cf. Clements (1916). Adams, using floristic and faunistic evidence, discussed postglacial dispersal, while Transeau (1905) described the forest centers of eastern United States, later dealing with the "prairie peninsula" and its relict outliers. Gleason (1923) explained the presence of xeric relicts in the Midwest as survivors of an appropriate postglacial climate, since replaced by a more mesic one.

In 1930, fifteen years after the appearance of von Post's technique of using fossil pollen to interpret changes in vegetation and climate, active use of this method began in the United States. Gleason's postulates were confirmed and general agreement between the European and American sequences was demonstrated. Applications to archeology and cultural history followed, with notable contributions in the arid Southwest and in Mexico, far beyond the limits of continental glaciation.

Meanwhile Chaney, in his study of plant megafossils from the Tertiary, has compared them with modern vegetation of related floristic composition. In that way he has been able to reconstruct the most probable associations of the earlier interval of geological time—an important contribution to paleoecology.

One of the most recent fields of ecological specialization is radiation ecology, sponsored by the Atomic Energy Commission. Among the results has been the finding of differing degrees of specific resistance among plants to damage from atomic radiation. Since one of the chief concerns in studying such effects is the problem of survival in the unhappy event of uncontrolled nuclear warfare, this is perhaps an appropriate place to mention still another development—population ecology. Although chiefly concerned with animal experimentation, it

does involve plants, particularly in studies of speciation and in the spacing studies of the agronomist.

As to animal ecology, it may fairly be said to be an offspring in this country of Cowles' work on plant communities and successions. Working at Chicago, Shelford found in this information a powerful key to animal communities, scarcely surprising in view of the ultimate dependence of animal life upon vegetation. Eventually this led to his joint publication with Clements of *Bioecology* (1939), thus launching a new rubric whose effects are increasingly evident in published field work. This broader viewpoint is essential in studies of nutrient and energy cycles and of ecosystems in general.

A still further step in the development of ecology in the United States is the full recognition of man as the dominant organism and a powerful geological and biological force. While this was clearly stated by Marsh more than a century ago and was increasingly evident to all plant ecologists comparing remnants of "natural" vegetation with cultural landscapes, it became a major concern during the prolonged drought of the 1930's which accompanied a period of grave economic disaster. Then and since it has been a responsibility of ecologists to inform the public of the consequences of irresponsible disruption of the great natural cycles which have sustained life by governing the production of renewable resources such as soil, water, food and fiber.

In this way what had begun as plant ecology, progressing into bioecology, came to absorb what social scientists had called human ecology, built largely on such analogies as succession, competition and community pathology. Today ecology connotes the entire spectrum of ecosystems, including man as a component rather than merely a spectator. In this respect ecology can no longer be regarded as a science in the conventional sense. It is instead a point of view that draws upon the best information it can get from whatever source to interpret the interrelations of all forms of life with their environments.

Exigencies of national policy have led to massive underwriting of scientific research. Under those circumstances those aspects of science which have the appeal of immediate benefits in terms of security and profit have been well nourished. This in turn has led to amazing development in the physical sciences and the experimental phases of biology. But the benefits of ecological knowledge can seldom be applied in the market place. They must operate for the general welfare and are often costly and deferred. For this reason support of ecological research has lagged.

One of the unfortunate effects of this situation has been the position of ecology in academic programs where it is essential to the education of future citizens, teachers and professional technologists, including engineers and planners. There are, however, hopeful signs that the present crisis in population increase and maldistribution may result in

greater academic recognition and financial support. Growing concern over pollution of air and water, more frequent use of the word "ecology" in public print, and recent generous support from foundations for ecological research and teaching are all causes for a degree of optimism.

Incidentally as a gauge of the role of the plant ecologist in this movement, specialists in that field were chosen in 1950 to head sections established in two universities to promote work in the study and conservation of natural resources. One of these plant ecologists has occupied an influential subcabinet position in the national government.

This account requires some reference to journals and organizations. In 1897 a popular journal of plant science *The Plant World* began publication. Gradually its articles began to be more professional in tone, with increasing emphasis on ecology, both descriptive and experimental.

Meanwhile at the 1915 meeting of the American Association for the Advancement of Science in Columbus, Ohio, a group of some fifty plant and animal ecologists formed the Ecological Society of America. The stated reason for this move was to pool the interests of these two groups of biologists, but the conservatism among older organizations, not to say scepticism as to the validity of this youthful field of inquiry was certainly a factor. Today the membership is growing and amounts to about thirty-five hundred, while ecology plays an increasing role in many applied fields and their official societies.

Eventually the growing mass of technical papers required an outlet and in 1919 *The Plant World* ceased publication and was replaced by *Ecology,* later supplemented by *Ecological Monographs,* both published by the Ecological Society. For an appreciation of the luxuriant flowering of a discipline and viewpoint that began quietly in two parvenu universities at the beginning of the century, an inspection of the contents of current numbers of these two journals is recommended.

None of the foregoing should be construed as minimizing the deep obligation of plant ecologists in the United States to their colleagues in the rest of the Americas or the Old World. Both in publication and personal contacts our debt is very great, not only to distinguished leaders in the English-speaking countries, for example Tansley, Godwin and Elton, but also to a host of continental scholars. There is after all a genuine fellowship among those dedicated to a study of the living landscape.

One of the landmarks in this fraternization was the International Phytogeographical Excursion of 1912 which crossed the North American continent. Among those from abroad were Tansley from England, Paulsen from Denmark, Stomps from Holland, Skottsberg from Sweden, Engler and Tubeuf from Germany, and Schröter, Rübel, and Brockmann-Jerosch from Switzerland. Old photographs show a genial

group, many of them with impressive beards and baggy clothing. Tradition preserves a remark by Brockmann-Jerosch who had eagerly looked forward to the sight of a virgin forest, "Ach, Gott, dies ist kein Urwald. Sehen Sie nur die Stumpfen!"

*　　*　　*

To the writer's regret the foregoing has been written on extremely short notice and under heavy pressure. He is acutely conscious of its limitations and profoundly sorry that the illness of his distinguished colleague and friend, Dr. Oosting, prevented his carrying out the original assignment. No names of those now living have been included, with the exception of a few surviving pioneers and those of the many brilliant younger workers whose bibliographies are pertinent to the discussion.

HORTICULTURE

George H. M. Lawrence

(*Hunt Botanical Library*)

No facet of the plant sciences reaches, and is the month to month concern of, so many persons as horticulture—especially ornamental horticulture. Horticulturists long have generated and supported the botanical explorations of such men as David Douglas, Robert Fortune, Reginald Farrer, E. H. Wilson, David Fairchild, Frank Meyer, to mention a few of the past. In the last decade or so, John Creech of the U. S. Department of Agriculture has been foremost among Americans to explore the eastern parts of the Orient for plants. Horticulture and botany are inextricably intertwined: the taxonomist identifies and names the garden plants, the physiologist concerns himself with their culture and the geneticist with development of new hybrids, while the mycologist and plant pathologist contribute to the healthfulness of the garden flora. Over the years, however, the propagation and distribution of new introductions from the wild, the selection of forms better suited to garden needs, and the production of new hybrids has come also from those nurserymen and seedsmen who are good plantsmen as well as merchants.

The beginnings of organized formal horticulture have their roots in Great Britain and on the Continent. They long antedate botany as the latter is defined today. Thomas Hyll [Hill] a London man of letters wrote one of the earliest works in English on gardening. John Parkinson of England was more horticulturist than herbalist. In France, André Le Nôtre, the third generation of noted horticulturists in his family, wrote ably on plant materials, and for French nobility designed many of the great gardens in the environs of Paris. In the same period De La Chesnée Monstereul wrote the first horticultural monograph *Le Floriste François* (1654), on the tulip. Jan and Gaspard Commelyn, of Holland, were both botanists and promotors of horticulture.

Hedrick's *A History of Horticulture in America to 1860* (1950) is the best single survey of this early period, an era when most botanists were also good horticulturists. Humphry Marshall, with John Bartram and his son William, did much in their time to make Philadelphia and environs the center of American botanico-horticultural activity. All three contributed materially to the collections and exportation of native American plants to England and thence to the Continent. David Hosack, M.D., founder of the Elgin Botanic Garden on the site of the

present Rockefeller Plaza in New York City, was also founder and first president of America's first, the New York Horticultural Society, founded in 1818. Antedating most of these is America's first nurseryman, Robert Prince, who in 1737 founded the Prince Nursery at Flushing Landing, Long Island. If not the very first nursery this was certainly the first in intercolonial and international trade. Succeeded by his two sons, William and Benjamin, William re-established his part as The Linnean Botanic Garden and, with his descendants, made it one of the country's chief garden centers for nearly a century. For decades the Princes were the leading exporters of American plants to Europe and the major importers from all parts of the world. The firm went out of business about 1865. Closely interlocked with the early period of American horticulture was the predecessor to today's landscape architect, the landscape gardener. An eminent exponent was Thomas Jefferson followed by Andrew Jackson Downing who in 1851 designed the grounds in Washington around the Capitol, the White House, and the old red brick Smithsonian Institution area. He broke from the formal French and Italian style to introduce the English or natural school of design.

Leaders of national impact of the last 100 years go back to the Scot, Peter Henderson who emigrated to New Jersey in 1843, was of local importance as a market-gardner, florist, and seedsman, and influential as the author of books on ormamental horticulture. His *Gardening for Profit* (1865) opened a new era in American horticultural literature (3rd and last edition 1888, reprinted in subsequent years). His domination was assured with his *Practical Floriculture* (1868), *Gardening for Pleasure* (1875), and *Henderson's Handbook of Plants* (1881) all represented by successive and revised editions.

Liberty Hyde Bailey, undisputed titan of American horticulturists from about 1888 until his death at age 96, was the first to elevate American horticulture from the level of a craft to a science. It was he, more than any other person, who made botany the basis of sound horticultural research, teaching, and practice. This he did at Cornell University by bringing in physiologists and chemists to investigate problems of plant culture and production, who stimulated—if not founded—the discipline of plant pathology through the work of mycologists. He was himself a plant breeder (not to be confused with the ubiquitous "pollen-daubers" of his day and this), who later encouraged geneticists to work with cultivated plants, and who (a former associate of Asa Gray, and acknowledged authority on *Carex*) was the first American taxonomist to apply the basic principles of taxonomy to the identification and nomenclature of cultivated plants. His international stature was achieved largely through his bold and radical development as an administrator of horticultural education at Cornell, and through his prolific activity as an author. Over a span of 81 years

he virtually flooded the English-language public with 66 titles of books (exclusive of multiple editions) and three encyclopedias (for bibliography see *Baileya* 3: 36-40. 1955). Of these publications three cornerstone works must be cited: his *Standard Cyclopedia of Horticulture* (1914, ed. 2, 1922, reprinted without change to 1968), *Manual of Cultivated Plants* (1924, ed. 2, 1949), and, with his daughter Ethel Zoe Bailey, *Hortus Second* (1940). The Bailey era witnessed the emergence of his lifetime contemporary, J. Horace McFarland who, although primarily a printer and publisher of horticultural books, periodicals, and trade catalogues, was also for half a century the leading American Photographer of horticultural subjects, and long the American authority on garden roses, and founder of the American Rose Society.

Among recent national leaders was Hedrick, longtime director of the New York State Agricultural Experiment Station at Geneva, who promoted and edited a series, national in scope, of the best American fruit crop monographs (see also George Darrow, p. 000). Among these are *Grapes of New York* (1908) followed by volumes on plums, cherries, peaches, small fruits, and vegetables (the last never completed). Although written largely by specialists on his staff, this literature, together with his historical surveys of American horticulture and agriculture, has made Hedrick's name indelible in any history of the science.

The importance of David Fairchild as a plant explorer is mentioned above. More important, however, are his establishment of, and contributions to the plant introduction section of the U. S. Department of Agriculture. Trained as a mycologist Fairchild was always a botanist and built his staff largely with botanists. While his own interest centered on the introduction and improvement of subtropical horticultural crops, it was he who brought to his section such men as Swingle, world authority on *Citrus;* Joseph Rock the country's best informed plant explorer and geographer of China and its western neighbors: Frank N. Meyer Dutch-born plant explorer of Manchuria, Mongolia, and China; and H. J. Webber specialist in subtropical botany and horticulture. Long interested in genetics, Fairchild was a founder of the American Breeders Association, superseded by the American Genetics Association in which he was equally active. His administrative work was to be given greater impetus by B. Y. Morrison, a man trained originally in landscape design, but whose later plant breeding investigations in *Rhododendron* (Azaleas) and *Narcissus* also brought him to national prominence.

American literature of tropical and subtropical horticulture, largely regional in application, is of national impact because of its inclusion of all house plants. Notable among recent encyclopedic works are three editions of the superbly illustrated *Exotica 3,* 1963 by Graf of the Julius Roehrs firm in East Rutherford, N. J. Commercial horticulture,

long an empirically treated facet of horticulture, came into its own as a science largely through the work and students of the plant physiologist Kenneth C. Post at Cornell, and especially with the publication of his *Florist's Crop Production* (1949). Concurrently, at Geneva, N. Y., and later at Michigan State University, Tukey was applying plant physiology, anatomy, and morphology to resolving problems in pomology, as were Heinicke and MacDaniels at Cornell, Victor Gardner at Michigan State Univ., Gourley at Ohio State Univ., and the doyen of Pacific coast botanists active in horticultural research, Chandler at the University of California, Los Angeles.

Development and urbanization of the West found American horticulture of the second half of the 19th century becoming specialized by climatic regions: (1) the temperate Northeast, (2) the moist subtropical Southeast and Gulf of Mexico region, (3) the more arid but subtropical Southwest, (4) the moist mild-temperate Northwest, (5) the Rocky Mountain States, and finally the (6) north central states characterized by a continental climate of seasonal extremes presenting its own problems. Horticultural development in each of these regions expanded in proportion to the vigor and progressiveness of its leaders. The century of American horticulture since 1860 has also prospered because the influence of its leaders extended beyond regional boundaries.

The Northeast:

The northeast quadrant of the country is the richest among the makers of ornamental horticulture in America, due in part to its longer establishment in point of time, to the leadership of its institutions, and to the strong public interest in ornamental horticulture. In all fairness, however, it is recognized that contemporary interest and activity on the Pacific coast stands second to none in the country. Philadelphia and Boston were the primary centers of the Northeast in the 1870's, with New York State and its great fruit growing and nursery industries becoming a strong third in the decades to follow. The Massachusetts Horticultural Society, founded in 1829, and early made famous through such leaders as the pomologist Wilder; the Cambridge horticultural journalist, nurseryman, and seed merchant, Hovey; and slightly later, the Wellesley, Massachusetts, banker, railroader, and philanthropist, Hunnewell. Perhaps more than any one man in that area Hunnewell established contacts for plant exchanges with botanists and plantsmen abroad, so that by the start of the 20th century his 40-acre estate possessed the country's largest collection of conifers, oriental ornamentals, and orchids—collections shared freely and the fountain-head for exotics distributed to much of America. Many of the initial plantings at the Arnold Arboretum, Jamaica Plain, Mass., established by Sargent, were of stocks from the Hunnewell estate. The

influential nurseryman Harlan P. Kelsey of East Boxford, Massachusetts, together with F. L. Olmstead, Jr., American Landscape architect (and son of F. L. Olmstead, designer of Central Park in New York City and many baronial estates), and Coville of the U. S. Dept. of Agriculture, were compilers-publishers of the well-intended but abortive effort to stabilize horticultural nomenclature through the production of *Standardized Plant Names* (1923, ed. 2, 1942). Recently there are Rehder's publications on woody plant identification based on his fifty years of taxonomic research at the Arnold Arboretum, and those by Wyman, also of the Arboretum, on the culture and uses of hardy woody ornamentals. Karl Sax, geneticist and plant breeder, devoted more than thirty years at the Arnold Arboretum to developing new and more suitable cultivars of woody plants for garden and landscape use. A gem within a dreary borough is the Brooklyn Botanic Garden under Avery where only one adventure is the Garden of Fragrances for the blind.

The Philadelphia and adjoining New Jersey/Delaware area antedates that of Boston as a horticultural center. Even before the Pennsylvania Horticultural Society was founded in 1827 nurseries and the publication of horticultural books were prominent there. Only one edition of M'Mahon's *American Gardener's Calendar* appeared during his lifetime, but eleven editions followed until 1857. Its success is attested by the imitations. The Osage orange trees next to Christ Church on Second Street in old Philadelphia are a living link with M'Mahon's nursery.

Notable among the earlier leaders of the post-1860 period is Thomas Meehan, a Kewite, who came to Philadelphia in 1848 where he was employed by the Scottish-trained nurseryman and horticultural writer, the elder Robert Buist. In 1853 he established Meehan's Nurseries and from 1859 to 1890 edited and published the influential *Gardeners Monthly* (title varies), superseded by *Meehan's Monthly* (ceased in 1902), and numerous popular works in botany and horticulture. Throughout the last one hundred years of Philadelphia horticulture there surfaces repeatedly the name Conrad. The first was A. F. Conrad cofounder in 1862 of the nursery firm, at nearby West Grove, Pennsylvania, of Dingee & Conard, later (1892) of Conard & Jones, and most recently (1924) of Conard & Pyle.

Under Pyle's aggressive leadership the firm became one of the world's leading producers and distributors of roses. Since then Sydney Hutton, longtime partner of the firm, acquired full ownership in 1951 with his son. Pyle, ably supported by fellow Philadelphian John C. Wister, promoted and established, through the American Horticultural Council (1947-1957) which he founded and of which he was first President, the concept of the unification of professional, commercial, and amateur horticultural organizations in this country under a single superbody.

On its merger with the American Horticultural Society the Council's united horticulture movement unfortunately lost impetus and disintegrated. The Philadelphia seedsman Henry A. Dreer, American-born of German immigrant parents, began his business in 1838 by collecting and selling seeds of native American plants. By 1863 this had expanded to include ornamentals and his was one of the city's leading seed firms. Soon thereafter with his son he added the nursery business in Riverton, New Jersey. Today, Philadelphia's leading seed firm, with farms in Colorado, California, and South America, is Burpee's, headed by W. Atlee Burpee, Jr.

Situated in Moorestown, New Jersey, across the Delaware River from Philadelphia, Wisconsin-born Rex D. Pearce established an international reputation as a seedsman specializing in unusual species and varieties of ornamentals. The firm, founded in 1933, reached its peak a decade and a half later, and scores of species assembled from far distant points throughout the world tested and first introduced by Pearce are now commonplace in the trade.

Close to the Delaware border, some thirty miles southeast of Philadelphia is Longwood Gardens, the one-thousand acre estate of the late Pierre S. du Pont, of which a part was planted as a private arboretum about 1800 by the brothers Samuel and Joshua Pierce. Continuing the strong botanical interest of his French forebears, du Pont began to assemble here in 1906 one of the country's great collections of ornamental plants, rich especially in conservatory-grown exotics distributed through houses that collectively comprise the largest American garden under glass. Today, under Seibert's direction it is one of the horticultural showplaces of the nation.

The favorable climate of the Finger Lakes region of New York made it a focal point, and Rochester the primary center, for associated seed and nursery activity. Undisputed leader in the period 1850-1880 was the horticultural editor and seedsman James Vick. His journals, including *Vick's Magazine,* and seed catalogues made his name a household word throughout the East, and contributed much to the popularization of plants and gardens. Contemporary was the Rochester firm of Ellwanger & Barry and its "Mount Hope Botanical and Pomological Garden." German-born and trained Ellwanger was the nurseryman and Irish-born Patrick Barry was the promoter. Barry edited horticultural journals, for example, *The Horticulturist,* wrote popular books, for example, *Barry's Fruit Garden,* that earned him the reputation of being a top-rank pomological author, and was responsible for an extended run of outstanding trade catalogues. More recently, the firm of Jackson & Perkins at nearby Newark, New York, became the world's largest rose producing nursery, under the direction of Charles Perkins, his three brothers, and the production of a growing stable of new cultivars bred by the internationally renowned Eugene Boerner. In

1967 the home base of the firm was moved to Newport Beach, California.

The headquarters for horticultural research by the U. S. Department of Agriculture is at Beltsville, Maryland. Of too great a magnitude to be treated here even summarily is the work there by scores of men of national repute. Emsweller may be singled out for his genetical and breeding work in the genus *Lilium;* Darrow, pomologist, specialist in breeding berry fruits and author of the monograph *The Strawberry* (1966); Creech active in plant introduction and plant exploration; and Skinner, a specialist in azalea and rhodendron breeding and Director of the U. S. National Arboretum. F. P. Lee gave national leadership to organized horticulture for the last two decades. Professionally a constitutional lawyer, he drafted much of the U. S. Plant Patent Bill which became law in 1930, was an organizer of the American Horticultural Council, officer of the American Horticultural Society, and for twenty years an important contributor to the development of the U. S. National Arboretum. His *The Azalea Book* (1958, ed. 2, 1968) is one of the most useful on the subject.

THE SOUTHEAST:

Impetus to horticulture in the southeast was given by the Belgian born nurseryman and botanist Berckmans whose Fruitland Nurseries in Augusta, Georgia, operated from 1852-1907. His nursery was testing garden, experiment station, and veritable botanic garden. Much of the history of plant introduction into the South is to be found in the catalogues of his firm. A near contemporary specialist on plants for subtropical Florida was Pliny Ford Reasoner who with his brother Egbert founded the Royal Palm Nurseries (1882) near Manatee, Florida. Still active this firm has introduced, tested, and distributed more tropical and subtropical exotics to the warmer parts of the lower South than any other. Soon after the turn of the century Hume, Canadian-born and Iowa-educated, made Florida the focal point of a sixty-year career as nurseryman (founder of Glen St. Mary Nursery), Professor of Horticulture at University of Florida, and later Dean of the College of Agriculture and Assistant Director of its Experiment Station. Among his contributions were *The Cultivation of Citrus Fruits* (1926), *Gardening in the Lower South* (1929), *Azaleas and Camellias* (1953), and *Hollies* (1953). In the Charleston area are Middleton Place Gardens begun 1741, visited frequently by the elder Michaux; and Magnolia Gardens associated with John Drayton, author of *The Carolinian Florist* (not published until 1943), adapted to Walter's *Flora Caroliniana.*

Contemporary Florida has been enriched by Menninger who introduced and reintroduced species of subtropical flowering trees and shrubs, by Montgomery in the establishment of Fairchild Tropical

Garden at Coconut Grove, by Robert and Catherine Wilson until recently at Coral Gables, for the large number of exotic ornamentals introduced through their Fantastic Gardens, and by the bromeliad specialist, Mulford Foster. The catalogues and other publications of the firm of George W. Park and his sons George and William are replete with introductions. Situated originally (ca. 1870) in Pennsylvania, the firm has been located since 1925 in Greenwood, S. C. In recent decades the Park family has also originated many cultivars.

More recently horticulture of the Gulf area benefited from the enthusiastic devotion of Ira Nelson at University of Southwestern Louisiana, Lafayette, for his ornamental plant breeding work, South American explorations, and the introduction of materials new to the area. O. J. Ward, a Kewite, works with alkaline clays and changeable temperatures of New Orleans to make Mrs. Edgar B. Stern's Longue Vue garden a public exhibit of local horticultural possibilities. Caroline Dorman's "Briarwood" near Saline, Louisiana, has been a determined experiment in the saving and growing of natives.

The Southwest and California:

Contributions of this area are of a different character from those made elsewhere. The West of 1860 was largely unexplored for its plants, but the intervening century has witnessed abundant botanical activity resulting in utilization of many plants native. The climate is more arid and at sea-level varies from subtropical to mild temperate, hence an entirely different and substantially new cultivated flora became part of that region's urban and suburban landscape. Before 1850, horticultural activities were associated with the missions of Spanish-Mexican occupation. With statehood in 1850 California began its fruit industry, first with grapes, then citrus and apricots, and nurseries promoted their offerings of fruit trees and ornamentals. Perhaps the first of the San Francisco Bay area was Walker's Golden Gate Nursery (1850-1871). Walker came to California in 1849, introduced many native trees to landscape use, and Australian plants, including what is believed to be the first introduction of *Eucalyptus* and *Acacia* (ca. 1856). With the railroad connecting east and west, in 1869 commerce— and horticulture—came into its own. In the 1880's in Santa Barbara, Elwood Cooper first established large plantings of *Eucalyptus,* that Australian exotic which perhaps more than any other modified the landscape from north of San Francisco to the Mexican border. Hosp was the first to establish plantings in considerable acreage for the florist trade, first to plant a substantial garden devoted wholly to cacti, and to introduce South African natives in variety. Fenzi, Pisa-trained Italian, gave California horticulture great impetus the decade before and after the turn of the 20th century. Known to his American associates as Dr. Francesco Franceschi, he settled in Santa Barbara in 1892,

remaining until 1912 when he left at the request of the Italian government for Libya, to establish an experimental garden. Franceschi's contribution to America's subtropical horticulture rests with the hundreds of species and cultivars introduced (or reintroduced) to California gardens and its landscapes. A prolific writer, his magazine articles promoted effectively his new introductions.

Armstrong Nurseries at Ontario, Calif., started about 1890 by John S. Armstrong pioneered in the testing and selection of suitable citrus cultivars, in the 1920's of olive cultivars, and later of avocados. Demands for land for commerce and urbanization gradually eliminated much of southern California's great orchards and so the firm shifted its emphasis to ornamentals, especially roses and camellias. Important to subtropical horticulture was Rixford who came to California in 1867, made a career as a newspaperman, but is remembered for his efforts leading to the introduction of the Smyrna fig from Asia Minor, and as an avid plantsman whose botanical collections of the native flora are in herbaria here and abroad.

California produced two great plantswomen, each of whom made her own indelible marks: Kate Sessions and Alice Eastwood. When asked how she got started in gardening, Kate Sessions, born on Nob Hill, San Francisco, in 1857, replied "I was always started; I grew up in a garden." Abandoning her school-teaching a few years after graduation from the University of California in 1885, she opened a nursery specializing in rare or unusual exotics, and a flower shop, in San Diego. The rent for her land was paid to the city in plants for the park department; thus was the San Diego landscape enriched. Constantly searching and importing from the world over, she introduced perhaps one hundred of the ornamentals now common to the region. A person of dynamic personality, she wrote well and effectively, traveled and lectured widely.

Alice Eastwood who came to California in 1892 was a self-taught botanist. Her contributions of more than sixty years to horticulture rested largely in her interest in the identification and naming of cultivated ornamentals, an interest further stimulated in later years by Orpet. Her knowledge of California's exotic flora led to a warm comradery between her and nurserymen, gardeners and designers, and specialists of such groups as succulents, fuchsias, acacias, and ferns, and devotees of Rocky Mountain and California wildflowers. Honors showered on her by garden clubs and horticultural organizations attest to the respect, esteem, and love with which she was so highly regarded. Unquestionably Alice Eastwood was *la grande dame* of American plantsmen. Among the many horticulturists associated with Miss Eastwood a few stand above the others. English-born and trained Orpet came to California in 1917, for a decade was Santa Barbara's Superintendent of Parks. He established the Japanese persimmon industry

in the state, and became a specialist in the botany and horticultural use of succulents—assembling a world-wide collection. His nursery of ornamental shrubs, vines, succulents, and bulbous plants was noted for novelties, introduced in large number to American gardens. Hugh Evans came to southern California from England in 1892, settled ultimately in Santa Monica where in the 1920's one would find in his nurseries the widest range of plants of pantropical origins tested and found suitable for the area. Often he collaborated with Orpet in the introduction of plants from Australia, Chile, and Brazil. Interested in public park advancement, he contributed materially of his new introductions, from San Diego to San Francisco and northward. For many years, when he needed critical determinations it was Alice Eastwood who came to the rescue. Also associated with Miss Eastwood for more than fifty years, was Scottish-born and trained McLaren who developed and was longtime Superintendent of Golden Gate Park, San Francisco. He came to the Bay area in 1887, and is remembered for his design and planting of the Panama-Pacific Exposition (1915) area. Of strong personality McLaren exerted wide influence on Pacific horticulture, especially in the establishment and planting of public parks and gardens. Another of Miss Eastwood's horticultural associates was Walther who came from Dresden, Germany, first to assist McLaren in 1913 in the establishment of the Panama-Pacific Exposition grounds, and later became Director of the California Academy of Science's Strybing Arboretum in Golden Gate Park. Walther's botanical contributions were largely from his study of Californian Crassulaceae, especially the genera *Dudleya* and *Echeveria*. Miss Eastwood's protegeé Elizabeth McClintock, a taxonomic botanist, while monographer of numerous woody plant genera dominant in Pacific Coast gardens, is active in the State's organized horticulture where her efforts, often in association with Mildred Mathias of the University of California, Los Angeles, have assured greater accuracy in the naming of its cultivated flora.

Sydney B. Mitchell, Librarian at Stanford by 1908 and at University of California, Berkeley, by 1911, was a leader in California horticulture for nearly forty years, during which time the genus *Iris* was his subject of an extensive nursery he maintained along with his library profession. A founder of the California Horticultural Society, he was the first editor of its *Journal*. Author of several gardening books, his last was *Iris for Every Garden* (1949).

Two men of the southwest responsible for much current knowledge of succulent plants are Hertrich and Harry Johnson. German-born and Austrian-trained Hertrich came to America in 1900 and to California in 1902 and three years later was employed by railroad magnate Huntington at his estate in San Marino, serving beyond Huntington's death in 1928 when the Henry E. Huntington Library and Garden was

established. Hertrich collected more than 9,000 mature specimens of cacti and other succulents making the San Marino garden outstanding. His *Palms and Cycads, their Culture in Southern California* (1951) provided one of the first horticultural monographs of those plants. In later years he focused on camellias from the study of which came his superb 3-volume monograph *Camellias in the Huntington Gardens* (1954-59). Demonstration plantings for the nurseryman and householder are particularly rich at the Los Angeles City and County Arboretum in Arcadia where its director, William Stewart, has concentrated on dry-climate species from South Africa, Australia, and India.

British-born Theodore Payne began his Southern California career in 1893 on the Modjeska Ranch in the Santa Ana Mountains where he was soon experimenting with the growing problems of the colorful natives from nearby canyons, e.g., *Fremontia, Lupinus,* and the Matilija poppy (*Romneya coulteri*). From his downtown nursery, purchased in 1903 from Hugh Evans, he dispersed natives and exotics to a worldwide clientele. Californians were stirred to an appreciation of native plants for gardens by his vacant lot displays and his five-acre plot in Exposition Park, Los Angeles. The Santa Barbara Botanic Garden, which features native annuals, Ceanothi, and other flowering shrubs of the chaparral as landscape subjects, was inspired by one of his many lectures, "Preserving the Wild Flowers and Native Landscapes of California." Payne's horticultural skills with the natives also flowered in the early years of the Santa Ana Botanic Garden.

Los Angeles born Harry Johnson, son of a nurseryman, has been closely associated with succulents for nearly fifty years. He was a plant explorer in tropical America during much of 1913-1921, working in the U. S. Department of Agriculture under Fairchild. In 1921 he purchased Sturtevant's nursery, re-establishing it in the Los Angeles suburb of Paramount as the Johnson Cactus and Water Gardens. His subsequent cactus collecting expeditions in the Andes introduced for the first time many genera and species hitherto known only to botanists. Many are the botanic gardens of other countries whose stocks came from Harry Johnson's collecting. Edmund Sturtevant, who introduced such tropical novelties as Cup-of-gold (*Solandra guttata*), had a fine garden in the old Cahuenga Valley district of Hollywood. Writing about gardens for the *Los Angeles Times* from 1898 until his death in 1954 was Braunton, landscape gardener, hybridizer, member of the City Park Commission, teacher, botanical collector, and contributor to Bailey's *Standard Cyclopedia of Horticulture*.

The Pacific Northwest:

With its moist moderately temperate climate this may well be the most garden-minded region of this country, strong in public support of parks and gardens, and the one area where the climate favors so many

of the wonderful introductions first made to British gardens from such Asiatic regions as Nepal, the Himalayas, Burma, and temperate India. While Portland is the Rose city of the Pacific region, Seattle and the Puget Sound area are renowned for the wide species range and excellence of the Rhododendron and Azalea collections. The region includes also the finest and largest of American bulb-growing establishments. It is the area first made famous by explorations for British gardens by David Douglas. One whose contemporary efforts have made the herbaceous flora in particular better represented in gardens of America and abroad is Carl English, Jr. For some forty years, conscious of conservation, he has distributed seed and plant lists, in a given year offering as many as 1,500 different natives. He and his wife and business partner, Edith, are well known also for their many articles in horticultural journals of America and Britain.

ROCKY MOUNTAIN STATES:

The miners and ranchmen who homesteaded the Rockies (or their wives and sweethearts) had little time for gardens, and it was only here and there that a treasured tree or nostalgic window box of geraniums gave promise of the horticulture that was to follow. Fruit trees were planted in Clear Creek three years after Denver was founded, but were carried away in a flood two years later. The fruit industry of Colorado's Western Slope was launched by Samuel Wade in 1881. The pioneer nurseryman Andrews landscaped General Palmer's wild garden in 1895 near Colorado Springs. His Rockmount Nursery at Boulder featured natives for over four decades. The discovery of the sensational mutation "Red Sunflower" at Boulder in 1910 by Wilmatte Cockerell, wife of Professor Cockerell, was publicized by Sutton and Sons in England, and Henderson and Burpee in this country. The Colorado Forestry and Horticultural Association, founded in 1884, took on new growth with the publication of *Green Thumb* (1944—) and the opening of Horticulture House in Denver. Native Rocky Mountain plants were perkily described during these years by Kathleen Marriage in her catalogues of Upton Gardens at Colorado Springs, plants later to flower in many a British rock garden.

THE NORTH CENTRAL STATES:

Shaw's Gardens at St. Louis, officially the Missouri Botanical Garden, exerted a leading influence in horticulture in the Middle West, especially in the era of the Louisiana Purchase Exposition of 1903. The Garden exhibits of orchids and water lillies have been outstanding and its *Bulletin,* later often enriched by Edgar Anderson's articles, popularized horticultural topics. The Garfield Park Conservatory at Chicago, described by Wyman as "easily one of the outstanding conservatories

in the country if not the world," features especially fine programs for the plant-conscious public. The Morton Arboretum at Lisle near Chicago, opened in 1922 by Joy Morton whose father founded Arbor Day, has focused on trees hardy at that latitude and outdoor plantings of flowering shrubs, hedges, ground covers, and old fashioned roses.

HAWAII:

Three names are important in tropical ornamentals in the islands, Hillebrand, Rock, and Neal. Rock who arrived in Honolulu in 1907 to regain his health, joined the College of Hawaii in 1911, was instrumental in planting exotic trees on the campus and about the city, and published a profusely illustrated *Ornamental Trees of Hawaii* (1917). Rock made an inventory of the plants growing in what is now Foster Botanic Garden, which garden had been planted by the physician Hillebrand whose home it was from 1851 until 1871. Hillebrand had also come to Hawaii afflicted with tuberculosis and regained his health there. He introduced many tropical trees from Java and elsewhere into the islands, as well as exotic birds. Marie Neal, born in Connecticut, trained in botany at Smith College, went to the Bernice P. Bishop Museum in Honolulu in 1920 as an assistant in malacology and on her own time worked assiduously as a botanist. From 1930 until her retirement she was the museum's botanist. Outstanding as gardening literature is her *In Honolulu Gardens* (1928) superseded by her *In Gardens of Hawaii* (1948, rev. ed. 1965). This is in a sense Bailey's or Rehder's *Manual* for the tropics because its coverage, while designed for plants cultivated in Hawaii, meets most of the needs of gardeners in tropical and subtropical regions elsewhere. It is an excellent handbook, albeit a heavy tome, for identifying tropical plants on 'round the world cruises.

*   *   *

American horticulture in the century since 1860 has, through the men and women who made it, contributed substantially both to garden inventories of the world and to literature about plants grown in them. Horticultural advancement in America has proceeded at a greater pace in the past fifty years than has the national economy. Through research there have been developed wholly new techniques: (1) in plant propagation, by such means as embryo culture, hormone growth stimulants, and environment control by the use of automated mist sprays; (2) in commercial florist crop production, by means of automated nutrient and water controls, elimination of seasonal flowering through controlled day-length, and in product marketing through advances in packaging at point of origin, and controlled temperature and humidity within the pack; and (3) in nursery crop production, through the use of plastic film mulches, disposable container-grown plants both for

lining-out stock and retail sales, and more adequate control of pests and diseases.

Gardening has changed with the nation's shift from a two- to a three-class society. The number of large private estates, with their corps of gardeners and their large conservatories, has become so reduced as to seriously affect the very existence of the great flower shows. The rapid expansion in size of the middle class, and the increase in leisure time for almost all, has resulted in a booming growth of interest in gardening. In most other countries commercial interests control most such societies: in America it is the amateur horticulturist who controls the administration of all special plant societies—now established for more than seventy genera or cultural groups. The mens' garden clubs of which more than two hundred are active in this country, and mostly affiliated with the Mens Garden Clubs of America, are made up of serious amateur horticulturists, maintaining and developing their own gardening interests. Womens' garden clubs number in the thousands and their more than three million members—affiliated with the Garden Club of America or the larger Federation of Garden Clubs—exert a considerable influence at state and national level, especially in behalf of public parks and gardens, conservation of national resources, and beautification of urban and suburban areas. Both mens' and womens' horticultural organizations, from local to national levels, actively support horticultural research (through grants and awards), and horticultural education (through scholarships and travel grants). There has been a substantial increase throughout the country of gardening interest in tropical and subtropical exotics, e.g., orchids, bromeliads, cacti, and succulents; an interest that has lead to rapid growth in the construction of home-attached and independent glass or plastic conservatories—equipped for the most part with automatic controls for watering, temperature, and humidity.

Horticultural literature produced in America each year for amateur, professional, and commercial interests is more than any one institution or library is known to possess. In 1968 there were published for these three interest groups more than 370 titles of strictly horticultural periodicals.

At no previous time in this country's history has horticulture involved so many persons, has the inventory of the number of genera, species and cultivars grown under cultivation been so large, or the influence of plants on the social welfare of a people been so significant.

MEDICAL BOTANY

Jerry Stannard

(*University of Kansas*)

As in classical antiquity so in colonial America botany and medicine were so closely related as to be practically indistinguishable. For nearly 250 years—from the time of the first English settlements on the Atlantic seaboard to the middle of the nineteenth century, the alliance remained entrenched, but with westward exploration and the development of systematic botany under Torrey, Gray, and their followers, botany succeeded in establishing itself as an independent study. Systematics in turn surrendered its preeminence as the various disciplines and subdisciplines of botany began to be included in the curricula of our universities. At the same time botany underwent refinements in the medical schools largely as the result of developments in chemistry and microscopy, and the resultant development of interest in new methods and techniques in pharmacology and pharmacognosy.

Properly speaking American medical botany began with the native Indians and their use of plants for medicine and ritual. With no physician within reach, the settlers often appreciated the attention of a *quioccos* or *wisocist* who purported to know the curative powers of the native plants. The identification of all plants used by the various tribes is not completely known because of linguistic differences and the reluctance of the Indian herbalist to reveal secrets. Fenton in "Contacts between Iroquois herbalism and colonial medicine" (1942) suggests some plants used. As early as 1588 Hariot reported that Sassafras was "of most rare vertues in phisick for the cure of many diseases," that tobacco "purgeth superfluous fleame & other grosse humors," and that "sweet gummes of divers kindes," one *Liquidambar,* were used.

Cornut (1635) emphasized medicinal properties of plants of eastern Canada, which he learned from French explorers or tested in the physic garden in Paris. Of Southeastern *Robinia pseudoacacia* he wrote: "Totius arboris follorum succus, aut ligni decoctum, vi summâ ad refrigerandum & astringendum est," thereby perpetuating an age-old medical tradition by applying it to uses of New World plants. A Virginia merchant wrote Robert Hooke in 1678 that the colonists preserved "spunk," the fungus *Fomes,* to burn as a moxa on open wounds to prevent infection, a use learned from the Indians. Although almost every writer on plants of the New World mentioned their curative properties, the Rev. John Clayton in a letter to Nehemiah

Grew in 1687 left the most extensive list of the plants which the settlers tried at the suggestion of the Indians. Hoffman in *Ethnohistory* (Vol. 11. 1964) suggests identifications for these. *Aristolochia serpentaria* was treasured into the middle of the nineteenth century for snake bite although *Smilax* roots and Wild ginger (*Hexastylis virginica*) and "Dittany" would do in an emergency, but Clayton was skeptical of their value; *Liriodendron* buds were an "opener of obstruction"; the leaves of *Magnolia virginiana* were used in fevers; *Angelica venenosa* "stopped the flux," loosened and purged (it "saved" Clayton's life in a serious illness) ; for what *Iris verna* was used Clayton could not learn from the Indians; *Solanum dulcamara, Polygonum spp.,* and several *Apocynums* were purgatives; *Cunila origanoides, Pycnanthemum incanum,* and other mints were used for salves, perspirants, and so on. Directions for use, whether infusion or powdered in a salve, were seldom specifically given.

Kitchen gardens were laid out in accordance with homeland plans. Within these gardens some of the popular medicinal herbs of the Old World were planted, many of which later became naturalized. Josselyn's account of New England (1672) "Of such Garden herbs (amongst us) as do thrive there, and of such as do not" contains about fifteen herbs which were used in contemporary English medicine, e.g., Burnet (*Sanguisorba minor*), Ground ivy (*Nepeta glechoma*) and Feverfew (*Chrysanthemum parthenium*) .

Kalm published as one of his papers on natural history in 1750 an account of "Lobelia as a sure remedy against venereal disease," and in his *Travels* (Eng. ed., 1770) devoted considerable space to remedies for ague or malaria, a frequent complaint among newcomers to the colonies. In the absence of Jesuit's bark (*Cinchona spp.*), the bark of the Tulip tree (*Liriodendron*) and Flowering dogwood (*Cornus florida*), the leaves of Cinquefoil (*Potentilla sp.*), and the powdered root of Water avens (*Geum rivale*) were held by the colonists to be beneficial. John Lyon in his "Journal," 1799-1814, recorded "receipts" he had obtained in the mountains of South Carolina for various ailments.

There were few medical school graduates in the colonies and early United States, and these usually practiced in the larger cities: medicine and pharmacy were widely practiced by home-trained herb and root gatherers or outright quacks. Herbals long the source of information on plant uses in Europe were of limited help in America although American plants were included as they were taken to England, yet it took a great deal of searching and some botanical knowledge to distinguish them among European plants. Benjamin Franklin recognized the need for a home herbal and so persuaded John Bartram to prepare an appendix "containing a description of a number of plants peculiar to America, their uses, virtues, etc." for a third edition of Short's

Medica Britannica which Franklin published in 1751. Benjamin Smith Barton edited for the use of physicians and students in the University of Pennsylvania the British Cullen's *Treatise of Materia Medica* (1812) adding notes on the uses of American plants. Barton had embarked on, though he did not complete, his own *Collections for an Essay toward a Materia Medica* (1798-1804), which was completed by his nephew, W. P. C. Barton as *Vegetable Materia Medica of the United States* (2 vols., 1817-18). Bigelow's *American Medical Botany* (3 vols., 1817-20), and his supplement to the first *U. S. Pharmacopoeia* (1820); *A Treatise on the Materia Medica* (1822); Rafinesque's *Medical Flora* (2 vols., 1828) with "above 100 figures . . . and remarks on nearly 500 equivalent substitutes"; and Carson's *Illustrations of Medical Botany* (1847) followed. At the request of the Confederate Surgeon-general, Porcher compiled *Resources of the Southern Fields and Forests, . . . a Medical Botany* (1863). Millspaugh's *American Medicinal Plants* (1884) is especially well documented with a useful index to the medical literature of this early time and particularly for knowledge of poisonous plants.

The Elgin Botanic Garden in New York, the first physic garden in the United States, was established by Hosack in 1801 for teaching his medical students, and in his catalog of the plants growing in the garden their therapeutic values were noted. W. P. C. Barton also established a physic garden for Pennsylvania Hospital and the medical school. Studies of the properties of native plants were encouraged by Rush and B. S. Barton, and several notable dissertations resulted: John Moore's on Fox-glove (*Digitalis purpurea*) (1800); Thomas Massie's on *Polygala senega* (1803); and William Downey's on Puccoon or Bloodroot (*Sanguinaria canadensis*) (1803). An infusion of *"Tilia caroliniana"* and Mistletoe (*"Viscum purpureum"*) were reported in Humphry Marshall's *Arbustum Americanum* (1785) to have been used with success in epilepsy. Wild ginger (*Asarum canadense*) was sought as a poultice for an open wound on one of Lewis and Clark's troopers in 1806.

As knowledge of American plants increased so did the science of chemistry and the analysis of plant materials.[1] Pharmacognosy, the science which deals with natural products as pharmaceuticals, may be said to have arisen in Germany and France in about 1806, and from that time on botany as a medical subject was gradually taken over by the pharmacists. The first United States *Dispensatory* (1833) compiled information on drugs not yet official. As successive editions appeared drugs of mineral origin and chemically prepared compounds began to replace botanics. The first United States *Pharmacopoeia* (1820) was

---

[1] Grateful acknowledgment is tendered to Dr. Vernon Haarstad of Tulane Medical School for his suggestions in this latter part of Medical Botany.

published by the Authority of the Medical Societies and Colleges with the purpose "to select from among substances which possess medical power, those the utility of which is most fully established and best understood; and to form from them preparations and compositions, in which their powers may be exerted to the greatest advantage." The sixth edition (1890), however, was the first to include tests and assays setting strengths and purity of drugs. Although the first edition included almost all of the plants thought by the Indians and early settlers to have curative powers, the only two native American plants accepted as official in the 1960 edition are May apple (*Podophyllum peltatum*) and Cascara sagrada (*Rhamnus purshiana*). The resin of *Podophyllum* was originally widely used as a cathartic though now it is only occasionally used in small amounts in other mixtures, and to destroy "venereal" warts, but its important use is now externally for certain papillomas. It has been found to destroy cancerous tumors in mice. Cascara sagrada was discovered in 1805 on the Columbia River by Lewis and Clark. It and *Rhamnus californica* had been used by the Mexicans and Spanish in California from early times, but it was not until 1887 that Parke-Davis investigated its use and marketed it. The literature on poisonous plants in the United States especially in relation to animals increased impressively between 1884 and 1900, and was stimulated by the Hatch Act (1877) which allocated federal funds for agricultural research. While only about one third of the poisonous plants of the country had been thoroughly tested by 1958, about two thirds had had some report.

Advancements in pharmacology and pharmacognosy whether in the United States or elsewhere depended not only on refined laboratory techniques, but on more exact knowledge of plant identities and structure as well. Development in the microscope and microtechnique led anatomists to make minute and detailed histological description of most of the medicinal plants known. In the twentieth century interest in the physiology of the medical plant has kept pace with the refinements of biochemistry, and has replaced the purely descriptive knowledge of drug plants. It has led to continued experimentation on physiological action and chemical composition. Isolation and synthesis of the active ingredients in specific plant organs and tissues has led away from the plants themselves to purified biologically active substances whether natural or synthesized. The isolation of morphine by Sertürner (1806), which was the first isolation of a pure alkaloid from a botanical source, was followed shortly by the isolation of many medically important alkaloids, principally by the French pharmaceutical chemists Pelletier and Caventou. The French physiologist and clinician Magendie realized the value of pure alkaloids in therapeutics, since dosage schedules could be controlled far more easily with pure drugs than with crude botanicals of varying composition. Many

of the substances such as cocaine and quinine, however, originally taken as part of the plant are still among the most important drugs. Youngken (1958) has admirably summarized the relationship between botany and modern medicine.

Late in the nineteenth century and on to the present, chemotherapy, based upon relationships between chemical structure and pharmacological action has led to emphasis on synthetic drugs such as sulfa drugs and barbituates. Many new drugs have been discovered by chemical alteration of natural products. Even though a drug might be completely the product of chemical synthesis, the design may have been derived from the structure of a botanically derived compound. Very few *new* drugs of any value have been the product of pure chemical empiricism.

The great bulk of the world's plants have not yet been analyzed for their potential utility, but research here as abroad progresses. With some drug plants such as *Cinchona* and *Rauwolfia* there are significant chemical differences among the several species, and so the botanists' function in identification has assumed even greater importance. In 1884 Rusby became the first regular botanist and pharmacognocist for Parke-Davis, and in 1885 he was sent to explore for *Erythroxylon* since the drug company wished to analyze cocaine and its uses. In 1917 with the world's cultivated quinine supply in the hands of the Germans, he and Pennell searched Colombia for native *Cinchona,* the trees growing as individuals scattered throughout hillside forests. World War II soon found the quinine supply in Japanese hands, and with United States soldiers incapacitated by malaria in the South Pacific, twenty-two leading American taxonomists were dispatched to explore especially Colombia and Ecuador in an emergency program to obtain the better yielding barks from among the several species. Differences in the chemistry of the plant products obtain among populations of wide-ranging species, for example, *Cinchona pubescens,* and there are chemical differences depending upon where a particular species is growing. Studies of the influence of growing season, soil conditions, moisture, methods of harvest and drying, and preservation, are in progress or needed to learn their effect on the potency and activity of a plant drug.

The search for medicinal plants, particularly in the tropics, has been advanced by such devoted workers as Siri von Reis Altschul, ethnobotanist at Harvard who has culled from over two million herbarium specimens 6,821 notes by the collectors on native medicinal uses of 5,000 species from 178 plant families. Schultes of Harvard is prominent among ethnobotanists engaged in ferreting out the native uses of precisely identified plants for re-examination of their biochemical and physiological characteristics. The Indian uses of curare (*Strychnos toxifera*) as a poison led to its analysis, and subsequently to experi-

mentation in the use of the drugs in electro-shock therapy. Curare and derived (synthetic) drugs find their major application in producing neuromuscular relaxation during surgical procedures. The use of the Creosote bush (*Larrea divaricata*) as a drug by Nevada Indians led to the discovery of a complex acid which prevents butter and lard from becoming rancid. Though the acid is now synthesized, it was investigation of the plant that led to its use. Licorice (*Glycyrrhiza glabra*) has recently been investigated here and abroad in relation to its ability to mimic the effects of the adrenal steroids, and its antispasmodic action has been investigated. *Aloe,* the "bitters" long in use, was used effectively, indeed, it was the most successful ingredient in healing burns suffered in the South Pacific from radiation fall-out in atomic-bomb experiments. *Viburnum, Aletris, Helonias,* Blessed thistle (*Cnicus benedictus*), *Piscidia,* and *Potentilla,* are examples of botanicals, extracts of which have become part of pharmaceuticals for antispasmodics.

Yucca, Agave, Strophanthus, and *Dioscorea* are among many genera under investigation for steroid sapogenins since they hemolyze red blood cells, and so are toxic to humans, but they have been used for centuries by the natives of Mexico, Central America, and Africa. They are being reinvestigated for their useful "precursor" value in synthesis of adrenal hormones, cortisone, and hydrocortisone. *Dioscorea* may well be the most important drug-yielding plant ever discovered. It is used as a source for the precursors of the steroid sex hormones used in oral contraceptives, cancer chemotherapy, etc. Mescaline from the cactus Peyote (*Lophophora williamsii*) is currently being investigated for its hallucinatory effects and, later, intense cerebral depression. The effects of other hallucinogenic plants such as Jimson weed (*Datura stramonium*), Nutmeg (*Myristica fragrans*), and Morning glory seeds (*Ipomoea* and *Convolvulus spp.*) are also currently under investigation as are the effects of Marijuana (*Cannabis sativa*).

From the parasitic fungus Ergot (*Claviceps purpurea*) long known in controlled doses as a uterine stimulant, but in excess as a cause of gangrene, have been derived a number of physiologically active alkaloids. One of these, ergotamine, produces constriction of the cerebral arteries, which function is thought to be responsible in relieving migraine headaches. Another ergot derivative, ergonovine, is the chief chemical source for lysergic acid, and is at last being intensively studied as a result of its unfortunate use by so many of our young people.

Because a knowledge of cellular biochemistry is so important in understanding the chemical effects of drugs on cellular function, and therefore on tissues, this has been a field for recent concentration of research. And since the essential metabolites of plant cells are generally the same for animal cells, it is often easier to begin demonstration of biochemical responses to a drug in certain lower plants such as yeast

and bacteria, and then to proceed to a study of their uses on laboratory animals. The development of the electron microscope has contributed materially here. Plant physiology and biochemistry unite for greater medical progress.

Microbiology is a field which has developed rapidly since the early 1900's with improved microtechniques. From isolation in pure culture, identification and description, have developed a knowledge of microorganisms as agents causing disease. Progress has been made in recognition of deep fungus diseases. Bacteriophages have been used in determining relationships between bacteria since many of them are species-specific and so give evidence of the validity of the bacteria species. Research on these is of continuing importance in regard, for example, to enterobacteria and the distinctions of such bacteria as typhus and cholera. Indeed the identification of genera of bacteria is based on both morphological and physiological properties, but further subdivision is physiological.

The role of certain microorganisms, especially among bacteria and fungi, as constructive agents, though known for centuries in the role of cheesemaking and other fermentation, has been a fertile field of botanico-medical research. An aspect of this function is used extensively in the synthesis of steroid hormones from botanically derived precursors. Often a microorganism will alter the structure of a compound in a specific manner which is difficult or prohibitive by known chemical methods. Fleming in 1929 had reported from England the anti-bacterial action of penicillin, and in 1940 its curative powers were demonstrated. Since that time the search for new antibiotics here as abroad continues. Penicillin-resistant strains of Staphylococci and Streptococci spur the search for therapeutic agents from other microorganisms such as *Streptomyces spp.* Agents for controlling viruses are prime desiderata.

Thus the trend in medical botany is ever toward specialization, whether of the taxonomist who must know which species the morphologist, cytologist, or physiologist is to investigate, or of each of these specialists, and the cooperation is interdisciplinary and international.

In fact, no history of science ever ends with a bang, or even with a whimper. The historians merely stop writing. Just where they stop, perhaps, is not too important. Always it has to be at some arbitrary point, because science itself never stops.

SIRKS AND ZIRKLE

SELECTED READINGS

The earliest history of botany for the United States was a bibliographical essay by a New York physician, Samuel Latham Mitchill (1814); it was soon followed by a short account by Rafinesque (1817 and again in 1836), and a review history anonymously, but quite certainly, by Cushing (1821)—all little known essays but well worth perusal. William Jackson Hooker (1825) presented the British viewpoint. Four American botanists wrote useful summary histories of varying accuracy: Gray (1841), Short (1845), Darlington (1849) who often recounts the story from personal acquaintance with the principals, and Brendel (1879-80). In the twentieth century synopses interspersed with comment are provided by Underwood (1907), Coulter (1926), and Murrill (1945). The widest possible variety of topics are covered in symposia: a review of the previous twenty-five years of botany edited by Gager (1935); one hundred years of natural history, by Kessel (1955); the French connections, by Leroy (1957); fifty years of progress as reflected in the growth of the Botanical Society of America, by Steere (1958); and the bibliographic side of the subject, by Buckman (1966).

Ainsworth, G. C. Historical introduction to mycology. *in* G. C. Ainsworth and A. S. Sussman, *The Fungi, an Advanced Treatise* 3 vols. Academic Press, New York. 1965. vol. 1, pp. 3-20.

Anderson, Edgar. *Plants, Man and Life.* Little, Brown, Boston. 1952. Reissued by Univ. Calif. Press, Berkeley. 1967.

Baker, Herbert G. *Plants and Civilization.* Wadsworth Publ., Belmont, Calif. 1965.

Baker, Kenneth F. A Plant pathogen views history. *in History of Botany.* Clark Memorial Library, Univ. California, Los Angeles. 1965. pp. 36-70.

Barnhart, John H. Some American botanists of former days. *Torreya* 9:241-257. 1909.

> One of the early sketches of the many he wrote on botanists for a bibliography of which see *Bull. Torrey Bot. Club* 77:163-175. 1950.

Barreau, J., editor. *Plants and Migrations of Pacific Peoples. A Symposium.* Bishop Museum Press. Honolulu. 1963.

Bassham, J. A. and M. Calvin. *Path of Carbon in Photosynthesis.* Prentice-Hall, Elizabeth Cliffs, N. J. 1957.

Historical approach throughout.

Berman, Alex. Striving for scientific respectability: some American botanics and the nineteenth century plant Materia Medica. *Bull. Hist. Med.* 30:7-31. 1956.

———. The Heroic approach in nineteenth century therapeutics. *Univ. Mich. Med. Bull.* 24:419-427. 1958.

Bonner, J. T. *Morphogenesis.* Princeton Univ. Press. Princeton. 1952.

Brendel, Frederick. Historical sketch of the science of botany in North America from 1635-1840. *Amer. Nat.* 13:754-771. 1879. 14:25-38. 1880.

Buckman, Thomas R., editor. *Bibliography and Natural History.* University of Kansas Libraries, Lawrence. 1966.

Papers by John C. Greene, Stearn, and Stannard concern American botany in various bibliographic contexts.

Cain, Stanley A. *Foundations of Plant Geography.* Harpers, New York. 1944.

Carlquist, Sherwin. *Island Life. A Natural History of the Islands of the World.* Natural History Press, New York. 1965.

Christensen, Clyde M. *Molds and Man.* ed. 2. Univ. Minn. Press, Minneapolis. 1961.

Constance, Lincoln. Systematics of the Angiosperms. *in* E. L. Kessel, *A Century of Progress in the Natural Sciences, 1853-1953.* 1955. pp. 405-583.

———. Systematic botany—an unending synthesis. *Taxon* 13:257-273. 1964.

Coulter, J. M. Development of botany in the United States. *Proc. Amer. Philos. Soc.* 66:309-318. 1926.

Crabb, A. R. *The Hybrid Corn Makers.* Rutgers Univ. Press, New Brunswick, N. J. 1947.

[Cushing, Caleb] Botany of the United States. *North Amer. Review* 13:116-118. 1821.

Darlington, William. *Memorials of John Bartram and Humphry Marshall.* Lindsay & Blakiston, Phila. 1849. reprinted with indices and intro. by J. Ewan, Hafner, New York, 1967.

See "progress of botany in North America," pp. 18-33.

Dawson, E. Yale. *Marine Botany, an introduction.* Holt, Rinehart and Winston, New York. 1966.

"A Hundred years of marine botany in the United States" pp. 295-301.

Dunn, L. C. *Genetics in the Twentieth Century.* Macmillan, New York. 1951.

Eames, A. J. *Morphology of the Angiosperms.* McGraw-Hill, New York. 1961.

Efron, Daniel H. *Ethnopharmacologic Search for Psychoactive Drugs.* Government Print. Off., Washington, D. C. 1967.

Esau, Katherine. *Plants, Viruses, and Insects.* Harvard Univ. Press, Cambridge. 1961.

> Excellent review woven with historical threads.

Ewan, J. Frederick Pursh, 1774-1820, and his botanical associates. *Proc. Amer. Philos. Soc.* 96:599-628. 1952.

———. Scientist on the Frontier. *in* J. F. McDermott, ed., *Research Opportunities in American Cultural History.* Univ. Kentucky Press, 1961. pp. 81-101.

———. William Bartram. Botanical and zoological drawings, 1756-1780. *Memoirs Amer. Philos. Soc.* vol 74. 1968.

——— and Nesta Ewan. John Lyon, nurseryman, and plant hunter, and his journal, 1799-1814. *Trans. Amer. Philos. Soc.* 53 (2) : 1-69. 2 maps, 2 figs. 1963.

———, and ———. John Banister and his Natural History of Virginia, 1679-1692. *Proc. Tenth Intern. Congress of Hist. Sci.* pp. 927-929. 1964.

Fernald, M. L. Some historical aspects of plant taxonomy. *Rhodora* 44:21-43. 1942.

———. Some early botanists of the American Philosophical Society. *Proc. Amer. Philos. Soc.* 86:63-71. 1942.

[Flanagan, Dennis, editor] *Plant Life.* Simon and Schuster, New York. [1957] Anthology of review articles mostly on plant physiology appearing in the *Scientific American* between 1949 and 1957, each introduced with historical comment.

Gabriel, M. L. and S. Fogel. *Great Experiments in Biology.* Prentice-Hall, Englewood Cliffs, N. J. 1955.

> Chronologies introduce each biological topic.

Gager, C. S., editor. Twenty-five years of progress in botany, 1910-1935. *Brooklyn Bot. Gard. Mem.* vol. 14. 1936.

> C. E. Allen on cytology; A. F. Blakeslee on genetics; H. A. Gleason on ecology; E. D. Merrill on systematic botany, etc.

Galston, Arthur W., *Life of the Green Plant.* Prentice-Hall, Elizabeth Cliffs, N. J. 1961.

Geiser, S. W. *Naturalists of the Frontier.* Southern Methodist Univ. Press, Dallas. 1937. ed. 2., 1948.

> For warmth and understanding this anthology of essays has not been surpassed. Fully documented.

Gleason, H. A. Contribution of the Torrey Botanical Club to the development of taxonomy. *Torreya* 43:35-43. 1943.

Goodale, G. L. Development of botany since 1818. *in* Edward S. Dana, ed., *A Century of Science in America, with special reference to the American Journal of Science, 1818-1918.* New Haven. 1918. pp. 439-458.

Goode, G. B. Beginnings of natural history in America. *Report U. S. National Museum* 1897, pt. 2: 357-466. 1901.

    Very useful for quick orientation of persons and events.

Grant, Verne. *Origin of Adaptations.* Columbia Univ. Press, New York. 1963.

Gray, A. Notices of European herbaria, particularly those most interesting to the North American botanist. *Amer. Jour. Sci.* 40:1-18. 1841.

    Reprinted in Sargent, *Sci. Papers of Asa Gray.* Boston and New York. 1889. vol. 2, pp. 1-21.

Hayes, Herbert K. *A Professor's Story of Hybrid Corn.* Burgess Publ., Minneapolis. 1963.

Hedrick, U. P. *History of Horticulture in America to 1860.* Oxford Univ. Press, New York. 1950.

    Useful but not wholly reliable for dates and details.

Hellman, G. T. The Tail of Taxonomy. *New Yorker* 40 (15) :41-77. May 30, 1964.

    Vignettes of members of the New York Botanical Garden staff.

Hershenson, Benjamin R. A Botanical comparison of the United States Pharmacopoeias of 1820 and 1960. *Econ. Bot.* 18:342-356. 1964.

Holmstedt, B. and G. Liljestrand. *Readings in Pharmacology.* Macmillan, New York. 1963.

Hooker, W. J. On the Botany of America. *Edinburgh Jour. Sci.* 2:108-129. 1825. Reprinted in *Amer. Jour. Sci.* 9:263-284. 1825. Review of Hooker's original article in *Boston Jour. Philos. and Arts* 2:503-504. 1825.

Hornberger, Theodore. *Scientific Thought in the American Colleges, 1638-1800.* Univ. Texas Press, Austin. 1945.

Humphrey, Harry B. *Makers of North American Botany.* Ronald Press, New York. 1961.

    Must be read with great caution for errors. Cf. *Rhodora* 64:186-190. 1962.

Jaffe, Bernard. *Men of Science in America.* Simon and Schuster, New York. 1944.

    George Sarton wrote a foreword.

Keck, D. D. Trends in systematic botany *in* Bentley Glass, ed., *Survey*

of Biological Progress, Academic Press, New York. 1957. vol. III, pp. 47-107.

> Well-written terse review of all phases of the topic.

Keitt, G. W. History of plant pathology. *in* J. G. Horsfall and A. E. Dimond, *Plant Pathology, an Advanced Treatise,* Academic Press, New York, 1959. vol. 1, pp. 61-97.

Kessel, Edward L., editor. *A Century of Progress in the Natural Sciences, 1853-1953.* Calif. Acad. Sci., San Francisco. 1955.

> Botanical papers by E. A. Bessey, Constance, Ewan, Florin, Good, Manton, Papenfuss, Steere, and Van Niel.

Kingsbury, J. M. Knowledge of poisonous plants in the United States. Brief history and conclusions. *Econ. Bot.* 15:119-130. 1961.

Knobloch, I. W. *Selected Botanical Papers.* Prentice-Hall, Englewood Cliffs, N.J. 1963.

> Abstracts. The original papers may therefore prove important.

Kreig, Margaret B. *Green Medicine. The Search for Plants that Heal.* Rand McNally, Chicago. 1964. Reissued by Bantam Books, without ills. 1966.

Kremers, Edward and George Urdang. *History of Pharmacy.* Lippincott, Philadelphia. 1946.

Large, E. C. *The Advance of the Fungi.* Henry Holt, New York. 1940.

Leroy, Jean F., editor. *Les Botanistes français en Amérique du Nord avant 1850.* Centre National de la Recherche Scientifique, Paris. 1957.

> 21 papers on enterprises and individuals (Bachelot de la Pylaie, Bourgeau, Delile, Lamare-Picquot, Michaux, Milbert, Rafinesque, Trècul, and others).

Lewin, R. A., editor. *Physiology and biochemistry of the algae.* Academic Press, New York. 1962.

McKelvey, Susan D. *Botanical Exploration of the Trans-Mississippi West, 1790-1850.* Arnold Arboretum, Jamaica Plain. 1956.

> Cf. *Rhodora* 59:181-184. 1957, for a few addendae.

Manks, Dorothy, editor. Handbook on origins of American horticulture. *Plants and Gardens* (Brooklyn Botanic Garden) 23 (3): 1-89. "1967" 1968.

> Excellent summary of all aspects of development of horticulture in the United States with much new material by twelve contributors.

Meisel, Max. *Bibliography of American Natural History. The Pioneer Century, 1769-1865.* 3 vols. Publ. by author, N. Y., 1924-1929. Reprinted by Hafner, 1967.

More than a bibliography—a historical tool.

Merrill, E. D. Botany of Cook's Voyages and unexpected significance in relation to anthropology, biogeography and history. *Chron. Bot.* 14:161-384. 1954.

Mitchill, S. L. A Concise and comprehensive account of the writings which illustrate the botanical history of North and South America. *Colls. New York Hist. Soc.* 2:149-215. 1814.

Murrill, William A. *Historic Foundations of Botany in Florida (and America)*. Publ. by author, Gainesville, Fla. 1945.

Orlob, G. The Concepts of etiology in the history of plant pathology. *Pflanzenschutz-Nachrichten Bayer* 17:185-268. 1964.

Ornduff, Robert. *Papers on Plant Systematics*. Little, Brown, Boston. 1967.

Paddock, W. and P. Paddock. *Famine, 1975*. Little, Brown, Boston, 1967.

Parris, G. K. *A Chronology of Plant Pathology*. Johnson & Sons, Starkville, Miss. 1968.

Pennell, F. W. Historical botanic collections of the American Philosophical Society and the Academy of Natural Sciences of Philadelphia. *Proc. Amer. Philos. Soc.* 94:137-151. 1950.

Prescott, G. W. *Algae of the Western Great Lakes Area*. Cranbrook Inst. Sci., Bloomfield Hills, Mich. 1951.

Rafinesque, C. S. Survey of the progress and actual state of natural sciences in the United States of America, from the beginning of this century to the present time. *Amer. Month. Mag. and Critical Rev.* 2:81-89. 1817.

————. Historical sketch. *in* his *New Flora of North America*. Part II. 3-15. 1836.

Remarkable catalog of names including local figures.

Raup, Hugh M. Trends in the development of geographic botany. *Annals Asso. Amer. Geog.* 32:319-354. 1942. Also available in Bobbs-Merrill Reprint series in Geography G-193.

Reed, Howard S. *Short History of the Plant Sciences*. Chronica Botanica Press, Waltham. 1942.

Especially useful for plant physiology.

Ricker, P. L. Sketch of botanical activity in the District of Columbia, I-II. *Jour. Wash. Acad. Sci.* 8:487-498, 516-521. 1918.

Documented.

Rodgers, A. D. III. *John Torrey*. Princeton Univ. Press, Princeton. 1942. Reprinted Hafner, New York. 1965.

Rollins, Reed C. Taxonomy today and tomorrow. *Rhodora* 54:1-19. 1952.

Rusby, H. H. Historical sketch of the development of botany in New York City. *Torreya* 6:101-111, 133-145. 1906.

Safford, William E. Useful plants of Guam. *Contrib. U. S. Nat. Herbarium* vol. ix. Government Print. Off., Washington, D. C. 1905.

> "One of the world's most fascinating volumes" (Edgar Anderson).

Sauer, Carl O. *Agricultural Origins and Dispersals.* Amer. Geog. Soc., New York. 1952.

Short, C. W. Sketch of the progress of botany in Western America. *Western Jour. Med. and Surg.* :324-350. 1845.

Singleton, W. Ralph. Early researches in maize genetics. *Jour. Heredity* 26:49-59, 121-126. 1935.

Sirks, M. J. and Conway Zirkle. *The Evolution of Biology.* Ronald Press, New York. 1964.

Smallwood, William M. and Mabel S. C. Smallwood. *Natural History and the American Mind.* Columbia Univ. Press, New York. 1941.

Smith, C. Earle, Jr. A Century of Botany in America. *Bartonia* 28:1-30. 8 pls. 1956.

> Samples of handwriting of 44 collectors.

Smith, Gilbert M. *Marine algae of the Monterey Peninsula.* Stanford Univ. Press, Stanford. 1944.

Stannard, Jerry. Early American botany and its sources. *in* T. R. Buckman, *Bibliography and Natural History,* pp. 73-102. 1966.

Stearn, W. T. Botanical gardens and botanical literature in the eighteenth century *in* Allan Stevenson, *Catalogue of Botanical Books* 2:xli-cxl. Hunt Bot. Library, Pittsburgh. 1952.

Steere, W. C., editor. *Fifty Years of Botany. Golden Jubilee Volume of the Botanical Society of America.* McGraw-Hill, New York. 1958.

> Forty authors review the progress of American botany. Includes a gallery of portraits of botanists awarded Certificates of Merit.

Stebbins, G. Ledyard. *Variation and Evolution in Plants.* Columbia Univ. Press, New York. 1950.

———. *Processes of Organic Evolution.* Prentice-Hall, Englewood Cliffs, N.J. 1966.

Stetson, Sarah P. Traffic in seeds and plants from England's colonies in America. *Agric. Hist.* 23:45-56. 1949.

Swain, T., editor. *Comparative Phytochemistry.* Academic Press, New York. 1966.

Taylor, William R. *Marine algae of the Northeastern Coast of North America.* ed. 2. Univ. Mich. Press, Ann Arbor. 1957.

True, A. C. History of Agricultural Experimentation and Research in the U. S. *U. S. Dept. Agric. Misc. Publ.* 251:1-317. 1937.
Bibliography of 327 titles and two indices.

Underwood, L. M. Progress of our knowledge of the flora of North America. *Pop. Sci. Mo.* 70:497-517. 1907.

Wagner, Warren H. Jr. Plant taxonomy and modern systematics. *BioScience* 18:96-100. 1968.

Walker, J. C. History of plant pathology *in Plant Pathology,* pp. 14-44. ed. 2. McGraw-Hill, New York. 1957.

Wallace, Henry A. and W. L. Brown. *Corn and its Early Fathers.* Mich. State Univ. Press, East Lansing. 1936.

Whetzel, Herbert H. *Outline of the History of Phytopathology.* W. B. Saunders, Philadelphia and London. 1918.

Zirkle, Conway. *Beginnings of Plant Hybridization.* Univ. Pennsylvania Press, Phila. and Oxford Univ. Press, London. 1935.

———. Early ideas on inbreeding and cross breeding. *in* J. W. Gowen, ed., *Heterosis,* Iowa State College Press, Ames. 1952.

———. Plant hybridization and plant breeding in the eighteenth century. *Agric. Hist.* (in press) 1969.

I am just receiving the first reviews of my Flora
of California, Pts. 1 and 2. The critics mostly
or even entirely confine themselves to verbal
slips, not touching general principles. It is, to be
sure, disconcerting enough to have such errors,
but after all the main thing is this: 'Has the
book got matter in it? Has it got stuff in it?
Is it meaty? Not is it *faultless*. A faultless book is
impossible. It is inevitable in the nature of the
human mind that such slips will be made, mistakes and blunders. But is the job a big one, is it
really worthwhile?'

WILLIS LINN JEPSON, FEBRUARY 3, 1910.

Index to Persons

AF411826